WILD

MUSHROOMS

WASHINGTON

Tom Cervenka

Publisher: Northern Bushcraft Publishing

Design: Tom Cervenka

Editing: Ong Kar Khalsa

Website: northernbushcraft.com

Contact: tom.cervenka@gmail.com

Photos: All photos in this book are reproduced with the generous permission of the copyright holder. See the photographic attributions section for a full list of interior credits and abbreviation codes.

Front Cover Photographs: main photo by Guy Kennedy; top-right by Tom Cervenka; center-right by Tom Cervenka; bottom-left by Alan Rockefeller; bottom-center by Ron Pastorino; bottom-right by Ron Pastorino.

Cover Sheet Photograph: Tom Cervenka

Back Cover Photographs: top-left by D. Kolegayev [*CC BY-2.0 via FLK*]; top-center by Tom Cervenka; top-right by username denAsuncioner [*CC BY-ND-2.0 via FLK*]; bottom-left by Jason Hollinger [*CC BY-2.0 via FLK*]; bottom-center by Alan Rockefeller; bottom-right by Alfred Crabtree [*CC BY-ND-2.0 via FLK*].

Disclaimer: Do not eat any wild mushroom without being certain of its identity. If uncertain, obtain an expert opinion on its edibility. Neither the author nor the publisher accepts responsibility for the reader's identification of mushrooms or the consequences of eating any of the mushrooms listed in this book.

Glossary: Terms are adapted from WordNet 3.0 Copyright 2006 by Princeton University. All rights reserved. This database is provided "as is" and Princeton University makes no representations or warranties, express or implied. By way of example, but not limitation, Princeton University makes no representations or warranties of merchantability or fitness for any particular purpose or that the use of the licensed database will not infringe any third party patents, copyrights, trademarks.

Contents

MUSHROOMS

Contents

Acknowledgements

I would like to acknowledge the generosity of Ron Pastorino, Tim Sage, Darvin DeShazer and Alan Rockefeller for their photographic contributions. Special thanks also to Jason Hollinger, and the many other talented photographers who made this guidebook possible. A complete list of photographic attributions is provided in the back of the book. I would also like to thank Ong Kar Khalsa for proofing, Jenny Dunklee for research, Petar Jeremic for map-making assistance, and Hector B. Herrera and Bianca Duran for their work as illustrators.

I dedicate this book to Ong Kar Khalsa, my source of steady, loving support.

Preface

This guidebook does not cover all edible mushrooms in the state of Washington, but rather focuses on sixty that are both common and relatively safe for novice mushroom pickers. A great deal of research and consideration has been directed towards selecting mushrooms that are most suitable. This guide assumes that the reader is not entirely familiar with how to identify mushrooms based on key characteristics.

Never has there been a more fascinating time to become involved in mushroom hunting or amateur mycology (the study of fungi). Thanks to recent advances in molecular biology, decades-old notions about the classifications of even the most common and familiar mushrooms are being challenged, refined or completely upturned.

In some ways, mycology is entering the golden age of citizen science, where anyone with a camera, some time, and the willingness to participate can make valuable contributions to our understanding of mushroom species and their ranges. The website **mushroomobserver.org** allows users to upload mushroom photos that they have taken in the field. In doing so, they not only receive human assistance in making identifications, but also add their observations to a worldwide database that allows anyone to examine which mushroom species are being spotted, where and when. The database contains over a million records that can be searched by species, geographic location, or observer. I encourage the reader to register for an account and add their own observations, or to check out some of the other online resources for mushroom identification listed in the External Resources section.

I do expect that the effort taken in this field guide to curate a selection of beginner mushrooms will pay dividends for the intended audience, and will provide the reader with the confidence to embark on some of their first mushroom hunting adventures. I am excited to present a field guide that serves as an informative and practical starting point for mushroom hunting in Washington.

Introduction

THERE are estimated to be over fifty thousand mushroom-forming species of fungus. Although the vast majority of mushrooms are not edible, several dozen species are widely collected for food. Mushroom hunting has been practiced for centuries; in many countries it is an established and popular family weekend tradition. The prospect of discovering edible mushrooms in the grass, at the base of a tree, or hidden among the leaves and moss of a damp forest fascinates young and old alike.

For the novice, learning about wild mushrooms can be intimidating. It's easy to become discouraged thinking that mastery over a wide variety of mushrooms is required before any can be collected and identified with confidence. Fortunately, that is not the case.

This book presents those wild edible mushrooms of Washington that are suitable for the novice picker. Sixty mushrooms have been carefully selected for inclusion. They represent the safest and easiest mushrooms for a novice to collect because they either have no poisonous look-alikes or have look-alikes that can be easily differentiated on the basis of a small number of telltale characteristics.

CAUTION

Do not attempt to match mushrooms by photographs alone. Photographs can fail to capture key characteristics. To make a positive species identification for a mushroom, you must read the species description.

The descriptions provided for each mushroom species are written in a manner that is concise but avoids mycological terminology. No assumption is made that the

Red-staining Russulas, fascinating but not edible

reader has a background in mushroom identification. Included for each species is information regarding its habitat, seasonality, and tips for cooking and preservation.

This guide is a lightweight and informative text, intended to be slipped into a backpack and carried into the field. May it become a well-worn and dog-eared companion in your mushrooming forays.

MUSHROOM BASICS

Mushrooms are neither plants nor animals, but rather belong to the kingdom Fungi. This broad category includes not only mushrooms but also yeasts and molds, such as the one from which penicillin is derived. Mushrooms are incredibly numerous and varied. Like the fruits of plants, they can assume an enormous range of shapes, colors and sizes. For example, a mushroom may take the form of a cup, a club, a bracket, a coral-like structure, or a puffball, to name a few.

To look at a mushroom in a grassy field or growing from the side of a tree, it may not be obvious that it is only a small part of a larger organism. The main body of the fungus lies beneath the ground or within an organic host such as wood or leaf litter. This unobserved part, the mycelial mass, consists of a network of microscopic, thread-like filaments called "hyphae" that

Basket of highly prized spring king boletes

form an extensive, interconnected, living mat called the "mycelium." The mycelium is the main body of the fungus that carries out the business of feeding and growth. In order to feed, the mycelium secretes digestive enzymes that break down the surrounding organic matter. The resulting dissolved nutrients (carbohydrates, amino acids, vitamins) are then absorbed through the walls of the hyphae. In the process of feeding, the fungus plays a role as the principal decomposer in nature.

A mushroom is referred to as a "fruiting body" because it is the reproductive organ

of the fungus. For both plants and fungi, a fruit is responsible for producing and dispersing reproductive seeds. In the case of a mushroom, the "seeds" are millions of microscopic cells called spores that are produced on the underside of the cap and then released. Wind, water, rain, and/or animals then act as carriers, transporting the spores to new locations. Each spore has the potential to initiate a new mycelial mass, but only a fraction will settle on the right type of substrate at the right time to germinate. The vast majority will instead dry out or be damaged by ultraviolet light.

The new mycelial growth that arises from a spore does not necessarily go on to produce its own fruiting bodies. Only mycelium produced from two spores of different mating types can produce mushrooms. Whether or not mushrooms are produced, the mycelium will grow and over time may become extensive. A mycelium can cover acres of forest and reach several feet beneath the forest floor. One honey fungus measuring 3.8 km across in the Blue Mountains in Oregon is thought to be the largest living organism on Earth.

The fruiting season is the time of year when mushrooms are produced. It may only last a few weeks or extend from early spring until winter. During this period the mycelium forms tissue that develops into an immature form of a mushroom, called a button. In some cases the mycelium will fruit year after year in the same place. Other mushrooms grow in fairy rings, so that each year as the mycelium expands the ring of mushrooms becomes larger.

The mycelium typically lives for many years, periodically producing fruiting bodies to ensure the dispersal of its spores. The abundance of mushrooms in any given fruiting season can vary dramatically from year to year. The factors controlling any given year's bounty are somewhat dependent on weather and location, but are not fully understood.

MUSHROOM NAMING

When referring to a mushroom species, two types of names are used: the common name and the scientific name. The common name is the one used in everyday parlance. Some examples of common names are **king bolete**, **chicken of the woods**, and **black morel**. Common names often refer to some characteristic of the mushroom or to a person. Not every species has a common name and some species have more than one. Although common names are widely used, they often differ across regions and languages, and are therefore prone to being inconsistent or even contradictory. Obviously, relying on common names can lead to confusion, which is why scientific names are necessary.

Scientific names are based on natural relationships that are considered useful for making universal classifications. The scientific name is written in italics and is formed in Latin in order to be language neutral. It consists of two parts. The first part is the genus name, which groups mushrooms together on the basis of a restricted set of common characteristics. The second part is the species name, which is unique from any other member of the genus. For example, the genus Boletus contains a number of closely related species, such as *Boletus edulis*, *Boletus mirabilis*, and *Boletus suillus*. In some cases, a species will exhibit variations that are too minor to be considered a separate species. Instead, a subspecies name is appended to the name with the designation "var," e.g., *Boletus edulis var. grandedulis* and *Boletus edulis var. albus*.

A complication in naming arises when there is disagreement among taxonomists. Until recently, morphological features of the fruiting body and spores were the basis for making classifications. Increasingly, however, classifications are based on DNA sequencing, which is a modern technique

in molecular studies. Recent findings in these areas are altering traditional classifications. Sometimes mushrooms enter a state of taxonomic flux when more than one scientific name is assigned to a species by different groups before a consensus has been established. Although the process of taxonomic reclassification might seem confusing, it is of little consequence to mushroom pickers.

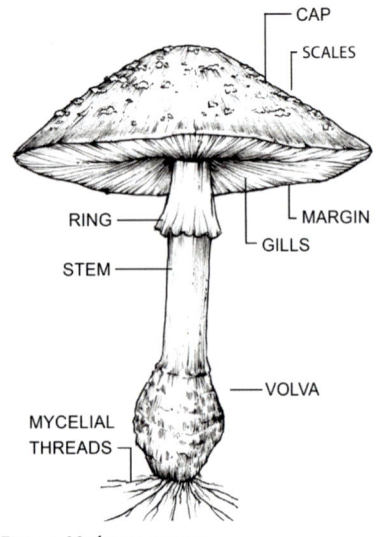

Figure 1. Mushroom anatomy

MUSHROOM ANATOMY

Mushrooms have a number of features that are useful for species identification. Gaining familiarity with basic mushroom anatomy is essential for recognizing these features and understanding how they can vary. This section introduces the anatomy of a typical (basidiomycete) mushroom and terminology for describing its features.

A mushroom begins life in an immature form of the fruiting body called a button. In the button stage, the cap or "pileus" curves downwards and inwards against the stem as it pushes its way upwards through the soil. As it grows, the cap gradually opens like an umbrella. In maturity, the typical mushroom consists of a cap and a stem or "stipe" that holds it up (*see Fig. 1*).

CAP

The cap is the upper the part of the fruiting body that protects the gills or "lamellae" on the underside, which are the site of spore production. In some cases the underside of the cap does not have gills but instead has spines or sponge-like pores. The central area of the cap is called the disc, and the area at the edge is called the margin. As the mushroom grows and matures, significant changes typically occur to the shape, color, and texture of the cap, disc and margin.

One of the most familiar cap shapes is "convex," (*see Fig. 2*) meaning that the cap is smoothly rounded, as seen in the **king bolete** (*p. 10*). When a cap is so rounded that it resembles half a sphere, it is said to be "hemispherical." The young **shaggy parasol** (*p. 25*), for example, has a hemispherical cap. In many species convex and hemispherical caps become flattened as they expand in maturity, and may even become upturned in age. An "umbonate" cap is one that has a softly curving,

humped protrusion at the center, such as in the **fairy ring mushroom** (*p. 33*). A cap that is triangular in its outline is termed "conical." For example, the **black landscape morel** (*p. 36*) usually exhibits a conical cap in maturity.

When a mushroom has a bell-shaped cap, it is termed "campanulate," whether or not the bell flares at the mouth. For example, the **inky cap mushroom** (*p. 22*) is campanulate. An alternate form is the "depressed" cap, where the center of the cap is lower than the rest, such as that of the **scaly hedgehog** (*p. 48*). A special type of depressed cap is the "umbilicate" cap, where the central depression is more severe, resembling a small hole. The cap of the **winter chanterelle** (*p. 5*) is umbilicate. Other mushrooms have a distinctly funnel- or trumpet-shaped cap, which is termed "infundibuliform." Because a mushroom's cap often changes in shape as it matures, it is not particularly useful by itself for mushroom identification.

The edge or "margin" of the cap can be classified in a number of ways, one such way being its general outline (*see Fig. 3*). If the margin is wavy, it is said to be "undulate." An undulate margin can be seen in **pig's ears** (*p. 6*) and the **pacific golden**

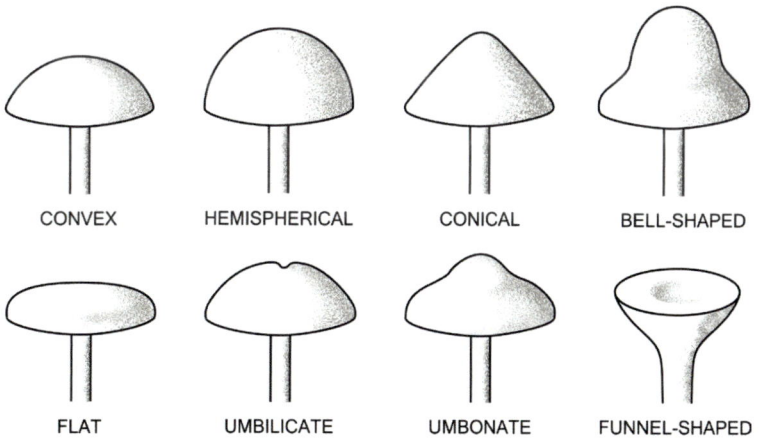

CONVEX   HEMISPHERICAL   CONICAL   BELL-SHAPED

FLAT   UMBILICATE   UMBONATE   FUNNEL-SHAPED

Figure 2. Cap shapes

chanterelle (*p. 2*). When the margin has ragged shreds of tissue adhering to and sometimes hanging down from the edge, it is termed "appendiculate." The hanging tissue is what remains of the veil, which is lost as the mushroom matures. The **king stropharia** (*p. 32*) sometimes has an appendiculate margin. If the margin exhibits radial grooves, it is termed "striate." If the margin is round-toothed or scalloped it is termed "crenate," whereas a "crenulate" margin is one that is very finely crenate. A "split" margin, as the name suggests, is one where split-like cracks interrupt the evenness of the edge. Some mushrooms, such as the **inky cap** (*p. 22*), have a margin that becomes ragged, which is termed an "eroded" margin. Another type is the "lobed" margin, meaning that it consists of two or more sections, or lobes, as seen in the **hedgehog mushroom** (*p. 46*). If the margin is even and regular, it is termed "entire."

Examining a cap in cross-section (*see Fig. 3*) reveals the margin's curvature. When the margin is bent upward, it referred to as "recurved" or "upturned." When the margin is bent inward to-

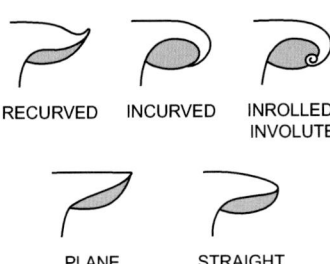

RECURVED   INCURVED   INROLLED/ INVOLUTE

PLANE   STRAIGHT

Figure 3. Margin curvature

wards the stem, it is referred to as "incurved," as seen in the **short-stemmed slippery jack** (*p. 14*). Many mushrooms have an incurved margin when they are immature. In some species the margin is inwardly curved and rolled up. This condition is referred to as an "inrolled" margin. If the margin curves neither inward nor outward it is said to be "straight." This margin type is not to be confused with a "plane" margin, which is one that is flush to a flattened cap.

Several surface qualities of a cap relate to its dryness, stickiness, sheen, and texture. When touched, the cap surface may feel dry, moist, or tacky. A tacky or sticky cap is termed "viscid" and it is often associated with a shiny lacquer, such as that of the **short-stemmed slippery Jack** (*p. 14*). A "glutinous" cap is one that has a distinct slimy layer, as seen with the **slimy spike cap** (*p. 29*). Viscid and glutinous caps often have debris adhering to the their surface, since their outer layer is sticky. In terms of surface sheen, a cap surface can range from dull and dry to glossy or silky.

The surface texture of the cap can be highly variable from species to species or over time. It may be smooth, wrinkled, pitted, cracked, scaly, striate (radially lined), warty, or some combination of the above. A cap with a smooth surface that lacks any cracks, wrinkles, or depressions is called "even." A split surface is referred to as "laciniated," a cracked one as "areolate," and a wrinkled one as "corrugate" or "rivulose"

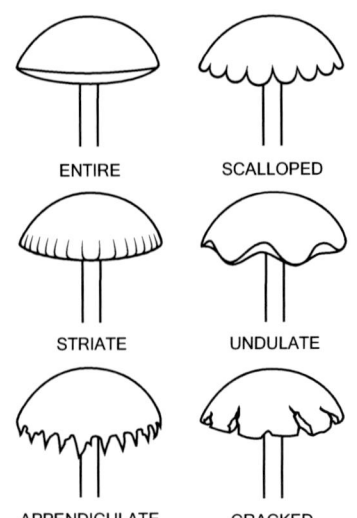

ENTIRE   SCALLOPED

STRIATE   UNDULATE

APPENDICULATE   CRACKED

Figure 4. Margin types

(see glossary for disambiguation). A cap may also be covered with fibrils, hairs, scales, granules, or fine powder, the last of which is termed a "bloom."

The coloration of a cap is one of its most obvious characteristics. Although it is a useful clue in species identification, it's common for color to vary over time or to be affected by environmental conditions. Prolonged exposure to direct sunlight, for example, can cause discoloration. Conversely, mushrooms that grow under leaves or in the shade can be unusually pale. Moisture and rain affect both color and translucency. Even in regular conditions a cap is not necessarily a single, uniform color; it may have different zones of color that blend into each other or may be mottled. It's also quite common for the central region of the cap to have different coloration than the margin.

The flesh within the cap can vary along a number of dimensions, including color, odor, taste, and consistency (soft, hard, firm, fragile, tough, woody, leathery, etc). In many mushrooms, the color of the flesh changes when exposed to air, in a reaction called staining. Staining can occur either immediately or after minutes or even hours have elapsed. In some species, the flesh produces latex when it is cut or broken. Latex is a milky or watery liquid that appears on the exposed area, seen in species of *Lactarius*.

Odor and taste, although subjective, can be used to further qualify the flesh of a mushroom. To assess odor, crush a small piece of flesh between two fingers. Mushroom odors are variously described as mealy, fragrant, fruity, indistinct, sweet, musty, and anise-like. Investigating the taste of an unknown mushroom is not recommended except in specific cases, such as when it is known to be some species of *Russula*. When tasting, a small sample should be briefly tested on the tongue and then spit out, never swallowed.

GILLS, PORES, SPINES

The gills are a series of thin, blade-like plates of tissue that occur on the underside of the cap. Not all mushrooms have gills. Some species instead have spine-like teeth, while others have small holes on the underside called pores. Cup fungi, morels, puffballs, and club fungi are examples of mushrooms that have neither gills, pores, nor spines.

Gills, pores and spines look radically different from each other but they facilitate the same reproductive function: spore production and dispersal. On a gilled mushroom at least some of the gills are full-length, meaning that they run all the way from the margin to the stem. In addition to full-length gills, some mushrooms have shorter gills called "lamellulae" that extend from the margin only partway to the stem. The lamellulae may occur in different lengths.

Depending on the species, the gills may be branching or unbranching. "Intervened" gills are those that have short veins running from the face of one gill towards another. A very important distinction should be noted between gills and "false gills." False gills differ from true gills in that they are not blade-like. Instead, they are vein-like or ridge-like folds of the undersurface of a mushroom. True gills, on the other hand, are separate structures that can be individually picked off. False gills are a key characteristic of chanterelles, such as the **pacific golden chanterelle** (p. 2).

Gills can attach to the stem in a number of different ways. When gills are broadly attached to the stem for all or most of their depth they are termed "adnate" (see Fig. 5). Gills that approach the stem (even very closely) but do not make contact are "free." The **deer mushroom** (p. 49), for example, has free gills. Alternatively, if the gills taper towards the stem so that they are only narrowly attached and are almost free, they are termed "adnexed." Adnexed gills occur in the **inky cap** *(p. 22)*. Sometimes gills are initially attached to the stem but pull away from it in maturity, leaving small lines at the top of the stem where they were previously attached. In this case, the gills are said to be "seceding." Finally "decurrent" gills are those that run down onto the stem, as seen in the **rosy gomphidius** *(p. 28)*. If descending only slightly onto the stem, the gills are said to be "subdecurrent." Oftentimes the type of gill attachment is variable even within a species or over time. For example, a species may have gills that are adnate to adnexed.

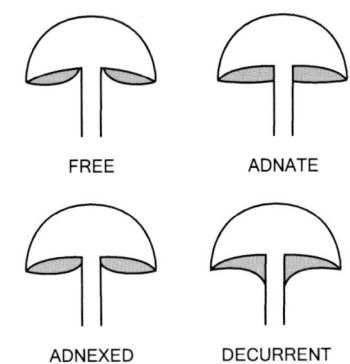

Figure 5. Gill attachment

The color of the gills is best determined by examining younger specimens before the caps have fully expanded and have been discolored by spores. Note that spore color cannot be ascertained simply by examining the color of mature gills, since they are often different. Gills may change in color after being bruised or as they age. For example the **shaggy mane** *(p. 24)* has gills that eventually decompose into a black, inky mass along with the rest of the cap.

Boletes and polypores are types of mushrooms that do not have gills. In place of gills they have a layer of vertical tubes, which constitutes the spore-bearing surface. The open ends of the tubes form a surface of pores on the underside of the cap. In boletes, the tube layer is relatively soft, spongy, and usually easily separated from the flesh of the cap, whereas in polypores the tube layer is tougher and more leathery or even woody. The characteristics of pores that are most important

for species identification are color, size, shape, color when bruised, and orientation (such as whether the pores are radially aligned).

Toothed mushrooms are those that have spines rather than gills covering the underside of the cap, for example the aptly named **hedgehog mushroom** *(p. 46)*. Such mushrooms use the spines as the site of spore production and dispersal. Noteworthy attributes of spines include their color, length, and whether they are blunt or pointy.

As with gills, if the spines or pores extend onto the stem they are said to be "decurrent." If they run down the stem only slightly they are termed "subdecurrent."

STEM

The stem of a mushroom, also called the "stipe," is the structure that supports the cap. Some mushrooms have a stem that is reduced to the point that it is not stalk-like,

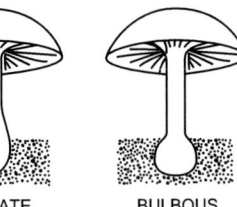
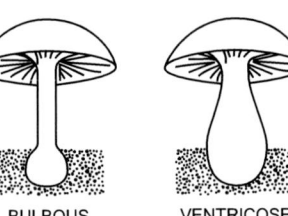

EQUAL   CLAVATE   BULBOUS   VENTRICOSE

Figure 6. Stem shapes

(e.g., **late oyster**, *p. 52*). In such cases the stem is termed a "rudimentary stem" or "pseudostem." Alternatively, the stem may be completely absent, as with puffballs and certain bracket fungi (e.g., **stump puffball**, *p. 43*). In mushrooms that do have a stem, it may attach to the cap at the center (central attachment), be off-center (eccentric attachment), or be attached at the side (lateral attachment).

How or whether a stem tapers is another noteworthy feature *(see Fig. 6)*. A stem that has parallel sides is said to be "equal." If the stem is enlarged below, forming a club-shape, it is termed "clavate." If enlarged in the middle, it is termed "ventricose." When enlarged only at the base, like a thermometer, it is termed "bulbous." In some species the stem is hollow, meaning that a channel runs longitudinally through the center. A stem that is not hollow is said to be "solid," and one that is filled with soft, cottony material is "stuffed."

The surface of the stem shares many descriptive terms with the cap (e.g., dry, shiny, dull, viscid, etc). As with the cap, the color of the stem may change over time or have complex coloration. Some surface characteristics, however, are specific to stems. For example the stems of some species have coarse channels or ridges that run longitudinally. Others have stems that are marked by small glandular dots ("punctate") or roughened with small pointed scales ("scabrous"). Another identifying trait of some mushrooms is the presence of a net-like pattern that resembles a mesh stretched over the surface of the stem, referred to as a "reticulum" or a reticulate stem. The **king bolete** *(p. 10)*, for example, has a reticulated stem apex.

Additional criteria used for classifying a stem relate to its dimensions, its strength, and its consistency of flesh. A stem may be unusually long or short in relation to the size of the cap, or unusually stubby or thin. A thin stem may be fragile or tough and wiry. It may be stiff or flexible, and when breaking it may split longitudinally or break in two. The flesh of the stem may stain after exposure to air, and may stain differently than the flesh of the cap.

VEIL

Many mushroom species have a protective structure known as a veil, of which there are two types: the universal veil and the partial veil. Some mushroom species have both, whereas others have only one or none. The presence or absence of a veil is a useful clue in species identification.

The universal veil, also known as the outer veil, is a protective, membranous tissue that envelops a mushroom when it is in the button stage. As the stem elongates and the cap opens, the mushroom stretches and breaks the universal veil, which often leaves remnants of tissue behind.

The remnants around the base of the stem form a cup known as a volva. The volva is a feature of *Amanitas*, some of which are deadly poisonous mushrooms. Other parts of the universal veil may remain on the surface of the cap, forming patches or warts. These warts can be a different color than the cap and are usually cleanly separable from it. The majority of mushroom species do not have a universal veil, and therefore lack patches, warts, or a volva. The easiest way to check whether a species has a universal veil is to examine a specimen young enough for it to still be intact.

In some species, the underside of the cap is covered by a piece of tissue called a partial veil. The partial veil does not envelop the entire mushroom but instead stretches from partway down the stem to the margin, covering and protecting the gills or tubes when the mushroom is young. The veil may be sheet-like (membranous) or so thin that it is cobwebby (filamentous). As with the universal veil, the partial veil tears as the cap grows, leaving remnants

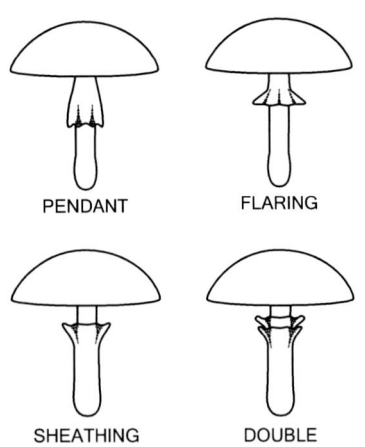

PENDANT   FLARING

SHEATHING   DOUBLE

Figure 7. Ring types

of tissue adhering to the stem, to the cap margin, or both. Veil remnants on the stem are referred to as a "ring" or "annulus." In some cases the annulus may become free and fall to the base of the stem. See **Figure 7** for ring types.

HABIT OF GROWTH

The proximity between individual mushrooms of the same fungus and how or whether they cluster together is referred to as the habit of growth. The habit of growth is an important characteristic that can aid in species identification.

A "caespitose" habit is one where fruiting bodies are joined together at their base, as seen with **pig's ears** *(p. 6)*. The term "connate" refers to a caespitose habit where the stems are joined together for some distance above the stem base. A "gregarious" habit indicates that fruiting bodies occur together in groups, whereas a "scattered" habit is one where they are widely separated (30-60 cm apart). If the fruiting body appears alone, the habit of growth is "solitary." It's common for a species to exhibit a range in its habit of growth, for example, growing scattered to gregarious.

FUNGAL ECOLOGY

Fungi can be classified according to the manner in which they obtain their nutritional requirements. Three major categories exist. Those that absorb nutrients from dead and dying material are "saprobic." Those that extract nutrients at the expense of a living plant or animal host are "parasitic." And those that grow in a mutually beneficial relationship with their host are "mycorrhizal." Understanding these relationships allows the mushroom hunter to associate mushrooms with specific hosts and habitats.

Saprobic fungi are the principal decomposers in ecological systems. As nature's recyclers, they obtain energy from digesting decaying organic matter such as plant litter, wood, and dung. The enzymes that a fungus produces determines the types of organic compounds that it can feed on. Their diet is typically specialized to some degree; for example, some may prefer hardwoods whereas others prefer conifers. A dead tree can act as a fungal host for several different fungi, either simultaneously or in sequence as it decays. Saprobic fungi can be grouped in two categories: brown rot and white rot. Fungi that are brown rot, such as polypores, are able to digest lignin. Those that are white rot, such as **oyster mushrooms** *(p. 51)*, are able to digest cellulose. Lignin and cellulose are the two primary components of wood.

Parasitic fungi extract nutrients at the expense of the plants, animals, and other fungi that they attack. Some parasitic fungi produce toxins that kill living cells, resulting in disease or death of the host. However the result is not always dramatic; some parasitic fungi rob nutrients without threatening the health of their host.

Mycorrhizal fungi are those that develop symbiotic (mutually beneficial) relationships with the living plants or trees that are their hosts. In nature, over 85% of all plants have mycorrhizal relationships with fungi. Mycorrhizal fungi operate by penetrating the roots of plants and then radiating out into the surrounding soil, forming a protective sheath and greatly increasing the roots' absorptive surface area. As the fungus mines the soil around the root for nutrients, it acts as a conduit for moving nitrogen and phosphorus into the plant. This arrangement benefits the plant by allowing it to grow in extremely poor soil. In exchange, the plant provides the fungus with amino acids, carbohydrates, and vitamins, which may be its only food supply. In many cases, there is a degree of specificity between the host and the fungus. Some are so specific that neither species can survive without the other.

PLANNING YOUR FORAY

Before embarking on anything more ambitious than a local walk, it's worthwhile to invest some time in planning your foray. A number of factors will influence its success. One of the first choices is selecting a location. Always obtain permission from

the owner or caretaker of the land before harvesting mushrooms, whether it is private land or federal land such as national parks, forests, or preserves, or state property, including state parks/forests. Failure to obtain permission can result in arrest for trespassing or fines for collecting without a permit.

The National Parks and National Forests of Washington provide many excellent areas for mushroom harvesting (see map on p. xvii). Other types of government land that present opportunities include:

- Washington State Parks
- Bureau of Land Management Lands
- Department of Natural Resources (DNR) Forest and Land Trusts (State Forests)
- Department of Fish & Wildlife Lands

Some types of government land, however, do not allow mushroom harvesting under any conditions. For specific rules and regulations as they apply to each district see the appendix *(p. 62).*

When deciding on a destination, consider that mushrooms grow best in shady, moist areas that have rich deposits of decaying organic matter. Good areas for mushroom hunting are often those near bogs, lake shores, and waterways such as rivers and creeks. Other good areas are damp, cool locations in meadows, ravines or foothills. Forests that have many fallen trees and tree stumps can be especially abundant with mushrooms, since the soil contains dead wood that retains moisture well. When foraging for a particular species of mushroom, target its known habitat. For example, you might focus your efforts near trees or plants that are known to be a mycorrhizal partner.

Avoid areas that may have been affected by industrial waste or other pollutants. Mushrooms can absorb chemical contaminants and toxic heavy metals from the ground. Do not pick in areas where herbicides or

---

RULES OF MUSHROOMING

Having a healthy attitude to mushroom picking means balancing ambition with caution. By following these rules, you can expect to have safe, enjoyable and rewarding experiences.

1. Only collect in permitted and uncontaminated areas
2. Do not eat a mushroom unless certain of its identity
3. Do not eat a damaged, overripe or very young mushroom
4. Do not eat a mushroom if it is not thoroughly cooked
5. Eat only a small amount when trying a new species

---

pesticides are routinely applied, such as lawns and golf courses. Mushrooms that are growing at busy roadsides should also be avoided, since they can contain heavy metals and other toxins that have been washed off the road by rainfall.

When travelling into the wilderness or other remote regions, designate a "check-in" person who remains behind and is aware of your destination as well as the planned route of travel, and your estimated time of return. The check-in person should be prepared to notify the appropriate authorities if you or your party fails to return within some agreed-upon time frame. An excellent way to find additional companions for foraging is by joining your local mushrooming club. If no club exists in your town or city, check for the nearest one and obtain a list of their planned forays, which will not necessarily be local.

Before setting out for a day of mushroom picking, make sure that you've packed the proper clothing, gear and supplies. Your clothing and footwear should be appropriate for the weather, temperature, and terrain. In the event that you become disoriented and lost in a remote area for any length of time, you may face life-threatening risks from exposure, hypothermia and dehydration. To protect yourself, carry an adequate water supply and water-resistant clothing, such as rain pants and a waterproof poncho. Check the weather forecast before departing and be prepared for rapid weather changes. The same rainfall that encourages mushrooms to grow also encourages the breeding of biting insects. Long sleeves and long pants will protect against pests and also minimize scratches from moving through the brush. Bring a hat in tick season to keep ticks out of your hair and check yourself for ticks at the end of the day.

Although expensive equipment is not required for mushroom picking, there is some basic gear that you will find invalu-

able. In order to carry mushrooms, bring a basket, cardboard box, woven bag, or other lightweight container that is rigid but not airtight. Don't bother with tupperware, plastic bags, or other airtight containers; instead, bring brown paper bags, waxed paper or paper towels for wrapping specimens and keeping them separated. An old toothbrush can be a handy tool for cleaning dirt and debris from specimens. Bring a writing pad if you plan to take field notes and bring white sheets of paper for making spore prints. A camera is not necessary, but it does allow you to quickly capture a mushroom's appearance in its natural environment.

A knife is handy for trimming and slicing mushrooms as well as prying them away from wood. Because you will be operating in moist environments, bring a knife that does not easily rust. A compass or GPS device and whistle should also be standard items in your kit. Without them, you may have difficulty navigating back to your starting point or alerting other members of your group if you become separated. It's also wise to pack a first aid kit that includes bandages and antiseptic ointment. To effectively protect yourself from most biting insects, carry bug repellent that contains DEET (diethyltoluamide). You may also want to pack along a canister of bear spray as insurance against unpleasant confrontations with bears. Use extreme caution when picking mushrooms in bear country and never approach or feed bears.

Choosing the right times to go foraging is as important as selecting the location. Fleshy mushrooms are composed mostly of water, so they are not abundant in dry years or after long periods without rainfall. At such times mushrooms might only appear in orchards, lawns, gardens, and other irrigated areas. In general, the best time for collecting is 2-5 days following heavy rains or sooner if the rain has persisted on and off for some time.

NATIONAL PARKS AND FORESTS IN WASHINGTON

1. Olympic National Park
2. Olympic National Forest
3. Mt. Baker Snoqualmie Ntl. Forest
4. North Cascades National Park
5. Okanogan National Forest
6. Wenatchee National Forest
7. Mount Rainier National Park
8. Gifford Pinchot National Forest
9. Umatilla National Forest
10. Colville-Kaniksu National Forest

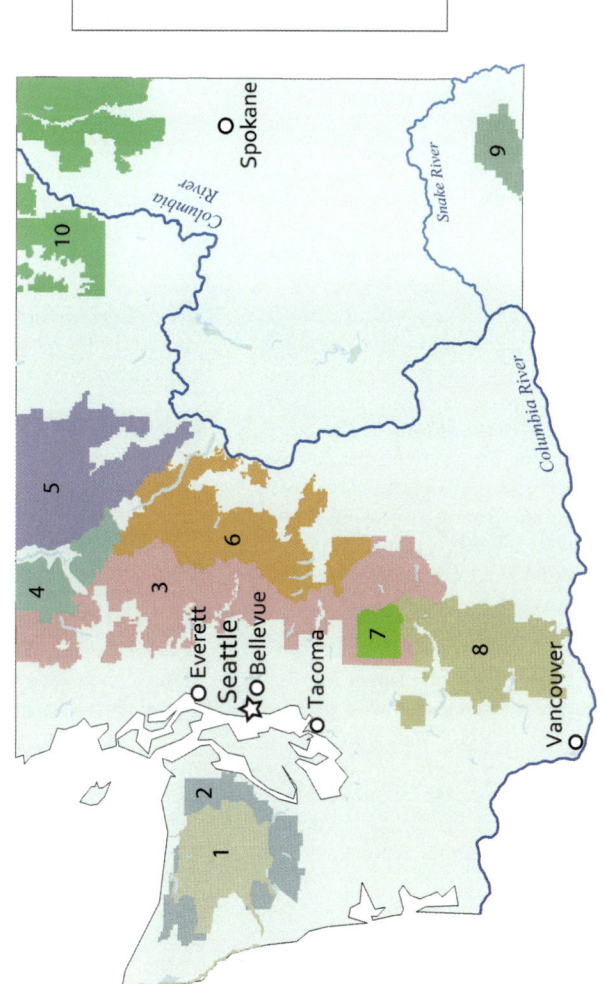

Mushroom fruiting is seasonal, meaning that mushrooms are produced at somewhat predictable times of the year. Because fruiting is influenced by temperature and moisture levels, however, it is idiosyncratic and difficult to accurately predict too far into the future. In the coastal areas and foothills of the Pacific Northwest, forays are often scheduled throughout the fall and into the winter. Some mushrooms such as morels are typically gathered in the spring, and boletes can be abundant in the summer. Inland regions tend to be progressively drier, and therefore less bountiful, though high elevations are productive after the snowmelt in late spring. In years when the frost comes early, expect the picking season to be abridged.

COLLECTING MUSHROOMS

One of the first considerations when collecting mushrooms is to ensure that they are in good condition. Pick mushrooms carefully one at a time, passing over any that have signs of over-ripeness, infestation, or damage. Avoid disturbing the forest floor by stripping away layers of moss, leaves or needles. As long as the substrate is not excessively disturbed, the process of harvesting a mushroom has no more ecological impact than picking fruit. After picking a mushroom, use a small brush to clean away any debris clinging to the cap or stem. Soil and grit particles are much easier to remove before they become lodged in the gills of other mushrooms.

When harvesting a familiar, edible mushroom, you may choose to field trim it by cutting away the base with your knife. Doing so will allow you to quickly check for any infestation or other damage. When harvesting unfamiliar mushrooms retain any important features at the base of the stem, such as the volva, since those features will assist you in making an identification.

Unless you take proper precautions, your mushrooms can disintegrate in their container as a result of being jostled or crushed. Having damaged specimens compromises your ability to separate, organize, and identify them. Even worse, when parts of different species mix together it can result in cross-contamination between edible and poisonous species. To avoid this scenario, create separate collections of mushrooms as you go. Each collection should contain only mushrooms of the same type. As part of a collection you may wish to include specimens at different stages of development, since some features that are useful for identification may only be present at a particular stage.

To prevent your collections from mixing together, keep each one in a separate basket or container, or wrap each collection in wax paper, which allows the mushroom to 'breathe.' Do not use plastic bags or tupperware to keep your collections separated, since airtight containers create a warm, moist environment where bacteria can thrive, causing the mushrooms to quickly become goopy and spoil.

A straightforward way of wrapping a collection is to lay down a piece of wax paper and place the mushrooms in a row down the middle. Roll the waxed paper around the mushrooms to form a cylinder, and then seal the ends by twisting them tightly. Wax paper has an advantage over regular paper in that it's less likely to soak through and come apart. When wrapping a collection of unfamiliar mushrooms for later study, it's useful to label each collection so that it can be associated with a set of field notes. Carry your collections in a basket or other rigid container. Since it doesn't require much force to crush a delicate mushroom, check your container from time to time to make sure that the heaviest items aren't crushing more fragile ones below.

Recording field notes and taking photographs of unfamiliar mushrooms pays dividends once you have returned from the field. To be comprehensive, your field notes should include any information that might assist you in making an identification, such as the substrate (e.g., ground, tree, stump) and the habitat (e.g., forest, lawn, meadow, moss, grass) in which the mushrooms were found, as well as the habit of growth and the types of shrubs or trees that grow nearby, which could indicate a mycorrhizal relationship. Taking clear photographs of both the top and underside of a mushroom is an effective way of rapidly capturing its characteristics. Photos can then be uploaded to online mushroom forums, along with field notes, to assist with identification (see external resources, p. 70).

When returning to your vehicle after a foray, be aware that a closed vehicle parked in the sun can become warm enough to accelerate spoilage. Upon returning home, process your mushrooms as soon as possible.

Before cooking or preserving your mushrooms, perform a close visual inspection of each one, looking for softness, sponginess, or foul odor, which are signs of over-ripeness. Slice through the stem and search for wormholes. Finding any of the above rules out the mushroom as being fit for consumption. For each collection, double check that all specimens are of the same type.

When dealing with an unfamiliar species, compare each mushroom against the full species description. Pay particular attention to the traits that differentiate it from its look-alikes. Very young mushrooms, such as those in their button stage or that still have a veil, should not be processed for food since they are relatively difficult to identify and cannot be used to make a spore print.

Figure 8: Spore print from a bolete.

MAKING SPORE PRINTS

One of the key characteristics of a mushroom is the color of its spores. The spore color can be a useful clue in identifying a mushroom and distinguishing it from any look-alike species, particularly when a comparison of macroscopic traits is not sufficient. When dealing with a new and unfamiliar mushroom, its always wise to ascertain the spore color. Individual spores are too small to see with naked eye but in mass deposit the color intensifies and becomes discernible. There are a number of ways to determine spore color, some more reliable than others.

Sometimes mushroom caps grow in overlapping clusters and the color of the spores becomes apparent as they accumulate on the cap beneath. The spore color can also sometimes be discerned by closely examining the mushroom itself, since spores tend to stick to the stem. These methods, however, are neither reliable nor accurate.

The recommended technique for determining spore color is to make a spore print. Spore prints are formed when a mushroom cap are arranged so that the spores are allowed to drop undisturbed onto a clean surface.

To make a spore print, cut the cap from a mature, fresh mushroom and place it with the gills (or other spore producing tissue) facing down onto a sheet of clean, white paper. Do not use colored paper because it will obscure the subtle difference between a pure white spore deposit and one that is off-white. It might seem that white paper would be a poor choice for recording a white spore print due to lack of contrast, however the deposit can nonetheless be detected by examining the paper from an angle. As an alternative to white paper, transparent glass can be used.

After placing the mushroom cap on the paper, cover it with an inverted cup or bowl to prevent any disturbance by air currents. It can take anywhere from thirty minutes to eight hours for a sufficiently thick spore deposit to form. If no spore print forms, it may be because the specimens are insufficiently mature, are sterile, have been physically damaged, or have been compromised by temperature or humidity. Note that a spore print cannot be obtained for mushrooms that do not discharge spores from their exterior (e.g., puffballs).

TOXICOLOGY

The majority of fleshy mushrooms are neither edible nor poisonous, but instead are considered inedible or of unknown edibility. Inedible mushrooms are those that are too small or unpalatable to be eaten. Just because a mushroom is palatable, however, does not mean it's edible. Some of the deadliest mushrooms reportedly have a pleasant taste. When a mushroom is said to have unknown edibility, it means that there are contradictory reports of its edibility or that its edibility has not been sufficiently established.

From time to time, bogus theories and folk wisdom surface on the topic of how to test for mushroom edibility. One common misconception is that mushrooms eaten by wild animals are safe for human consumption. Another is that edibility can be determined by examining the color that a mushroom turns when being cooked. A third claims that mushrooms are edible if the skin easily peels away. Do not give credence to such rules of thumb.

The only reliable method of establishing edibility is accurate identification, correct preparation, and sampling small amounts before consuming larger quantities. There is no substitute or shortcut to accurate identification. Even after identifying a mushroom as an edible species, it's wise to assess how your own body will react to eating it for the first time. Edibility often depends just as much on the person as on the mushroom. Some mushrooms are regularly consumed by many people but nonetheless trigger adverse reactions in sensitive individuals. For this reason it is impossible to recommend any species as being safe for everyone. To gauge your own body's reaction, begin by sampling a small amount of well-cooked mushroom – under a quarter cup for adults and under a tablespoon for children.

A complicating factor in determining edibility is that individual mushrooms can vary in their toxicity, even if they are of the same species. Genetic differences between strains is one factor that can account for such variance. Another factor is geographic location and environmental conditions. For example, a mushroom collected from a polluted area may be toxic while the same species collected from another region is edible.

Whether or not an adverse reaction occurs can also depend on interactions with foods that were consumed either before or after eating the mushroom. For example the **inky cap mushroom** *(p. 22)* contains a toxin that arrests the metabolism of alcohol, resulting in the rapid onset of nausea and vomiting when they are consumed together.

Figure 9. Some poisonous mushrooms, across from top: (a) Clitocybe dealbata, (b) Amanita smithiana, (c-d) Amanita phalloides, (e) Galerina marginata (f) Hypholoma fasciculare (g) Amanita muscaria var. alba, (h) Paxillus involutus (i) Gyromitra esculenta

Another factor affecting toxicity relates to the under cooking of mushrooms. The cooking process itself makes some mushrooms such as **morels** and **honey mushrooms** safe to eat by destroying toxins that are otherwise present. Cooking does not offer any protection against spoilage, however. A mushroom that has become infested or over-ripe can potentially cause sickness. This relatively common scenario is the leading cause of mushroom-related calls to the poison center.

Mushroom toxins can be classified in a number of ways. The following three categories divide toxins by their effects on different parts of the body: cellular poisons, nerve poisons and gastrointestinal irritants. Cellular poisons as a group cause the most severe damage and are the most likely to result in fatality. Mushrooms that contain cellular toxins include certain species of *Amanita*, *Galerina*, *Gyromitra*, and *Lepiota*. The effects of cellular poisoning may not become apparent for a relatively long period (6 to 48 hours). Symptoms include abdominal discomfort, diarrhea, dehydration, vomiting, jaundice, low blood pressure, fast heartbeat, and abnormally low body temperature.

Another group of toxins is the nerve poisons, which work by disrupting nerve impulses. These toxins are found in certain species of *Amanita*, *Clitocybe*, *Inocybe*, *Panaeolus* and *Psilocybe*. Symptoms include abdominal pain, vomiting, diarrhea, and dehydration, drop in blood pressure, decreased heartbeat, muscle spasms, convulsions, paralysis, and coma or even death. The effects of poisoning usually appear within minutes, but the onset can be delayed for several hours.

The third category, gastrointestinal irritants, is a "catch all" for toxins that cause stomach upset. The majority of poisonous mushrooms fall into this category. Symptoms typically occur within a couple of hours. The severity of symptoms is highly variable across individuals. The same mushroom that causes severe gastrointestinal irritation for one person can leave another completely unaffected.

If you believe that you may have eaten a poisonous species, contact the **Washington Poison Center:**

(800) 222-1222

ORGANIZATION OF THIS GUIDE

The species accounts within this book are arranged so that each page describes one mushroom species, or one species complex (group of related species). The presentation order is loosely based on genus for the sake of continuity, but no separation is made for higher orders of classification. Mushrooms may be included that require special preparation before eating, as well as mushrooms whose primary value is medicinal rather than culinary.

Each species account includes:

A. **Scientific and common names**
B. **Color photographs**
C. **Bullet-point key characteristics**
D. **Detailed description**
E. **Measurements**
F. **Look-alikes**
G. **Occurrence**
H. **Culinary use and other comments**

A. Scientific and Common Names

The scientific name of the species is provided at the top of each page, immediately above the first photograph. The common names are then listed in order of descending frequency of use, as sampled across both scientific and non-academic literature. The most frequently used common name is presented as the header. The remaining list of common names is separated by commas, with a forward slash (/) denoting interchangeability of words within a common name.

B. Color Photographs

Great care has been taken in selecting the color photographs for each species in this guide. Photographs are selected for their effectiveness in conveying the typical or most representative form of the fruiting body, and for highlighting as many of the key features as possible. Whenever possible, photographs have been included that reveal details of both the upper surface and undersurface, as well as the stem and the habit of growth if possible.

C. Bullet Point Key Characteristics

The bullet point list of key characteristics highlight some of the most important identification features of the species, many of which are not apparent from the photographs alone. The bullet point list does not provide every key characteristic discussed in the detailed species description. To establish a positive match, the detailed species description must be consulted.

D. Detailed Species Description

The detailed species description provides an account of each part of the mushroom. To facilitate comparisons, mushroom features are described in the same order from one species to the next: first those relating to the cap, then the spore-bearing surface, then the stem.

When attempting to determine whether a specimen in hand matches one of the species descriptions, it is essential to compare it against the complete list of characteristics. Unless all the characteristics match, the reader should not assume that the mushroom is edible. A poisonous species may closely resemble an edible species except for one or two crucial differences.

Attempting to identify mushrooms by matching them to photographs rather than to the full species description can result in misclassifications.

E. Measurements

In order to improve readability, all measurements are provided in a separate section from the detailed species description. Values signifying typical measurements are written as a hyphen-separated range, and values signifying extreme measurements are provided in parentheses.

F. Look-alikes

The look-alikes section provides guidance for differentiating the species being described from similar looking species. For every look-alike that is listed, the species name and edibility is noted along with a list of some of the most prominent differentiating features.

A look-alike may be edible, inedible, of unknown edibility, or poisonous. When an edible look-alike is mentioned, it may or may not include a reference to a full-page detailed species description. The reader should not attempt to identify an edible look-alike species for consumption unless it is listed in the table of contents.

G. Occurrence

The occurrence section provides information regarding the mushroom's picking season, habit of growth, and its preferred habitat.

H. Culinary Use & Other Comments

This section describes the culinary value of the species, including tips for cooking and preservation. Also included may be the mushroom's medicinal properties and miscellaneous notes.

The Mushrooms

Pacific Golden Chanterelle

- resembles a yellow funnel
- underside has ridges, not gills
- ridges are forking
- ridges extend onto stem

DESCRIPTION: The fruiting body is characteristically funnel-shaped. When young the **CAP** is convex with inrolled edges. As it matures, the cap flattens and then becomes depressed or bowl shaped in the center (but not perforated). The edges of the cap either remain inrolled or become upturned and irregularly wavy. The cap is normally orangey-yellow but varies with growing conditions. In wet weather it is more vibrant. In dry weather it develops a thin, greyish to brownish covering of dry scales. The surface texture is suede-like or slightly roughened from scales. The **FLESH** within the cap is firm and whitish.

The **UNDERSIDE** of the cap has deep, forking, ridges that extend down onto the stem. These ridges are called "false gills," since they are not blade-like, but instead are vein-like folds on the surface. The undersurface ranges in color from whitish with yellow/pink tones to pale orange-yellow. The **SPORE** deposit is pale yellow. The **STEM** is fleshy, solid (not hollow), and gracefully tapers downwards. It is similar in color to the cap or paler.

SIZE: Cap is 2–15 cm broad; stem is 4–6 cm long and up to 1½ cm thick at the apex.

LOOK-ALIKES: A number of gilled mushrooms can appear similar, including those that have acquired a funnel-ish shape from becoming upturned and exposing their gills in age. Such mushrooms differ by having "true" (i.e., blade-like) gills rather than forking, vein-like ridges. The poisonous *Omphalotus olivascens* differs in that it has true gills and grows in clusters where the stems are fused. The poisonous *Paxillus involutus* differs in that it has a brownish cap, has true gills, and has a brown spore deposit. The probably edible *Chroogomphus tomentosus* differs in that it has true gills and the spore deposit is smoky-black. The probably edible *Hygrophoropsis aurantiaca* differs in that it has true gills and a white spore deposit. The edible and similar-looking *Cantharellus cascadensis* is difficult to distinguish in the field; it differs in that the cap is often brighter yellow and the stem may not be tapered. See also *Cantharellus roseocanus (p. 4)*.

OCCURRENCE: In the late summer and fall this mushroom appears, growing either alone or in small groups on the ground where each remains separated. It grows in association with conifers.

COMMENTS: This chanterelle is a choice edible that is highly sought after, despite causing rare allergic reactions. It is a long-lived mushroom and resistant to infestation. Its has a faintly sweet smell that is fruity or apricot-like and becomes more noticeable when the mushroom is dried.

White Chanterelle

- resembles a white funnel
- underside has ridges, not gills
- ridges are forking
- ridges extend onto stem

DESCRIPTION: This eye-catching mushroom is the albino of the chanterelle family. The fruiting body is irregularly funnel-shaped. The **CAP** is initially convex with incurved edges. In maturity it becomes broadly convex to flat, develops a central depression, and the edges become uplifted and wavy or lobed. Its color ranges from ivory white to cream and discolors to yellowish-brown or orangey when bruised or handled. The surface texture ranges from smooth to nearly felt-like. When old it develops shallow cracks as well as fine, flattened scales and pale yellow-brown stains. When dried, the entire fruiting body becomes dark orange. The **FLESH** within the cap is firm and cream-colored. It inconsistently and slowly discolors to dull yellow when cut. The flesh has a peppery taste when a small piece is sampled, then spit out.

The **UNDERSIDE** of the cap has deep, forking ridges that extend well onto the stem. These ridges are called "false gills," since they are not blade-like, but instead are vein-like folds on the surface. The **SPORE** deposit is white to cream. The **STEM** is fleshy, solid (not hollow), and tapers downwards. It is similar in color to the cap. When bruised, it discolors to dull yellow-brown, especially near the base.

SIZE: Cap 5–10(15) cm broad; stem is 2–5 cm long and half as thick.

LOOK-ALIKES: A number of poisonous and inedible white mushrooms can appear similar, including those that have acquired a funnel-ish shape from becoming upturned and exposing their gills in age. The inedible *Lucopaxillus albissimus,* for example, can appear similar when old. Although its gills extend into the stem, they differ in being "true" (i.e., blade-like) gills rather than ridge-like, false gills.

OCCURRENCE: In the fall and winter this chanterelle appears on the ground either alone, scattered, or in groups where each remains separated. It grows in association with conifers, especially **Douglas-fir** and **hemlock**. Look for it on duff-covered ground in old growth forests. It is common in coastal and- montane areas throughout the Pacific Northwest.

COMMENTS: Like other chanterelles, this mushroom has a mildly sweet odor reminiscent of apricots. It is considered a choice edible and has a mild, pleasant flavor when cooked.

Rainbow Chanterelle

syn. Summer Chanterelle

- cap is pinkish when young
- cap is paler than the orange underside
- underside has ridges, not gills
- ridges extend onto stem

DESCRIPTION: The **CAP** is initially convex and lobed with edges that are rolled-under. In maturity the cap flattens and the surface often develops a lumpy or mangled appearance. The center of the cap often becomes shallowly depressed over time. In age, the edges of the cap become progressively thinner, more ruffled, wavy and irregular. When young, the cap is pale yellow and is often frosted with a pinkish bloom, especially near the edges and more so when wet. In maturity the cap becomes pale yellow to egg-yolk yellow or orange. The edges of the cap retain their pinkish/buff bloom even in age, when the rest of the cap becomes dingy and fades to nearly whitish. When bruised, the cap discolors very slowly to brownish, if at all. The surface is smooth or may have a few matted scales in age. When wet, it becomes tacky. The **FLESH** within the cap is whitish and soft-fibrous. When cut, it slowly turns yellowish, then ocher-brown.

The **UNDERSIDE** of the cap does not have "true" (i.e., blade-like) gills, but instead has blunt, vein-like ridges known as "false gills." The false gills are well-spaced, wavy and cross-veined. They run from the edge of the cap and extend onto the stem. The color of the young gills is peachy or apricot orange, then bright orange or yellow-orange in maturity. The gills are characteristically more intensely colored than the cap. When bruised, they slowly discolor to brownish if at all. The **SPORE** deposit is pale orange-yellow. The **STEM** is solid (not hollow) and variable in shape, but is generally short and stocky. It may taper in either direction and is colored like the cap or the false gills. The stem surface is smooth when young, then wrinkled, bumpy or grooved when old. When cut, the flesh within the stem slowly discolors brownish, if at all.

SIZE: Cap is 4–10(15) cm wide; stem is 2–6 cm long and 1–3 cm thick.

LOOK-ALIKES: A number of gilled mushrooms can appear similar, including those where the cap becomes upturned in age, exposing the gills. Such mushrooms differ by having "true" (i.e., blade-like) gills rather than forking, vein-like ridges. See also *Cantharellus formosus (p. 2)* and its poisonous look-alikes.

OCCURRENCE: In the summer and fall, this mushroom appears either alone, scattered, or growing in loose clusters on the ground. It is often found in older forests, where it may be buried in the duff under spruce and two-needle pines.

COMMENTS: This mushroom is considered excellent, particularly if collected when young and free of debris. It has a slightly peppery flavor and a sweet, fruity odor reminiscent of apricots.

Winter Chanterelle

syn. Yellowfoot, Funnel Chanterelle
Trumpet Chanterelle, Winter Mushroom

- cap has hole into stem
- underside has ridges, not gills
- stem is yellow and hollow
- spore deposit is cream

DESCRIPTION: When the picking season for most other mushrooms has ended, it's time to start looking for these under-sized late-bloomers. The **CAP** is initially convex with incurved edges. It soon flattens and the edges become wavy and uplifted or arched. In maturity the cap develops a central depression that becomes a funnel-like hole extending into the stem (i.e., "um-bilicate"). The cap is yellowish to brown and may exhibit faint radial streaks. The **FLESH** within the cap is thin, fragile, and pallid yellow.

The **UNDERSIDE** of the cap does not have "true" (i.e., blade-like) gills but instead has blunt, vein-like ridges, known as "false gills", which are the spore-bearing surface. The ridges are well-separated, fork frequently, and extend partly onto the stem. The underside ranges in color from yellowish grey to violet grey and is paler than the cap. The **SPORE** deposit is creamy/white. The **STEM** is smooth, slender, and is roughly equal in thickness throughout its length. When young it is stuffed with loose tissue, then becomes hollow, some-what flattened, and puckered. The common name **yellowfoot** refers to the stem's color, which is yellow to dull yellowish orange.

SIZE: Cap is 2–5(8) cm wide; stem is 3–8 cm long and 3–8 mm thick.

LOOK-ALIKES: A number of other mushrooms are superficially similar in appearance (such as also having an umbilicate hole in the center of the cap) but do not match all the key charac-teristics. For example *Chrysomphalina chrysophylla*, of unknown edibility, differs in that it has blade-like gills, has a cap that is not umbilicate, and has a spore deposit that is yellowish. The edible *Hygrocybe cantharellus* is superficially similar in appearance but differs in that the cap size is smaller (6-20 mm broad), the cap color is pale orange to reddish orange, and the gills are blade-like rather than ridge-like.

OCCURRENCE: In the winter and spring these common mushrooms emerge, either scattered or loosely clustered together on the ground, sometimes in large numbers. They grow near streams or in wet, boggy areas in association with conifers, particularly **western hemlock**. Look for them on well-decayed logs that are covered in moss, on rotting wood chips, or in the thick duff of the forest floor.

COMMENTS: These mushrooms have a subtle, pleasant flavor and an odor similar to chan-terelles. They work well when sautéed or cooked into soups and can be dried and ground into flavoring powder. When preserved by drying they reconstitute fairly well.

Pig's Ears

syn. Violet Chanterelle

- resembles pig ears from the side
- underside is dull violet, wrinkled
- stems are fused and not hollow

DESCRIPTION: This species typically fruits as a tight cluster of pinkish fruiting bodies that resembles pigs' ears, at least to those with imagination. When young the fruiting body has the shape of a flat-topped club that is continuous from the upper surface ("cap") to the stem. Over time the **CAP** becomes sunken in the center and the edges begin to develop asymmetrically. One side becomes greatly uplifted to form a lopsided funnel. The edges of the funnel are wavy and irregularly lobed. The **CAP** is dull ocher to tan and has a texture that is smooth to slightly felt-like. In age it fades to yellowish brown and develops tiny, scattered scales. The **FLESH** of the cap is firm, somewhat rubbery, and white to pale buff.

The **OUTER** (under) side of the fruiting body, which is the spore-bearing surface, is lilac- to purple-tan in maturity, then fades to yellowish tan when old. It is covered in shallow, blunt, forking wrinkles that descend almost to the base of the stem and are not easily scraped away. The **SPORE** deposit is light to dark brown. The **STEM** is short, thick, solid (not hollow) and tapers downwards. It is attached either centrally or laterally and is continuous with the rest of the fruiting body. Its color ranges from lilac to purple-tan near the top and whitish at the base and discolors to pale brown when handled or bruised. The base is typically fused with needles or duff. The flesh within the stem is white to pale pink and does not discolor when cut.

SIZE: Cap is 2½–10(15) cm broad; stem is 1–5(10) cm long and 1–3 cm thick.

LOOK-ALIKES: The not-recommended *Turbinellus floccosus* differs in that the stem is hollow, the cap has scaly orange patches, and the underside has pale ridges that are not lilac. Although eaten by some, for others it causes severe gastrointestinal upset. The not-recommend *Turbinellus kauffmanii* differs in that the mature fruiting body is vase-shaped and the depressed cap is brown with prominent scales rather than being smooth or felt-like.

OCCURRENCE: In the late summer and fall this common mushroom sometimes appears so-alone but more often occurs either in fused pairs, in small clusters that are fused at the base, or in overlapping clusters that sometimes form arcs or fairy rings. It grows on humus, often near well-rotted logs and can be found growing in partnership with conifers such as **spruces** and **firs**.

COMMENTS: This mushroom is a good edible that has a firm texture and a flavor that is somewhat similar to the chanterelle. It should be eaten with caution, however, since it causes severe gastric upset in some individuals. When harvesting, cut the stem above where it is fused to duff.

Blue Knight

syn. Blue-capped Polypore

- cap is bluish
- underside of cap has tiny, white pores
- pores descend down the stem
- often grows in fused clusters
- spore deposit is white

DESCRIPTION: This mushroom consists of one or more fused fruiting bodies that arise from a common base. The **CAP** is either circular, kidney-shaped, or has an irregular outline. When young it is convex with incurved edges. In maturity it flattens out or develops a central depression and the edges become wavy or lobed. The upper surface is initially blue grey or sometimes blue green. As it ages it exhibits stains or patches that are salmon to dingy ochre, then eventually becomes dingy ochre overall. Blue tones persist longer in protected areas where adjacent caps overlap. The texture of the upper surface is smooth to suede-like. It becomes wrinkled over time and cracked when dry. The **FLESH** within the cap is thick, tough, and rather brittle, breaking cleanly when fresh. The flesh within the cap is creamy white and discolors to pinkish buff when dried.

The **UNDERSIDE** of the cap consists of tiny, sponge-like pores that descend onto the stem. The pores, which are sometimes difficult to see, are round to angular. They range in color from white to pale cream when young, then sometimes become bluish and finally salmon or red when old. The pores are the openings of a tube layer - which is the spore producing surface - and cannot be easily peeled away from the cap. The **SPORE** deposit is white. The **STEM** is short, stout, and may be branched. It is solid (not hollow) and is either equal in thickness throughout its length, or may be wider at either end. It attaches to the cap(s) either centrally, off-center, or laterally. The surface of the stem is white or is tinged the same color as the cap, and has a smooth surface. The flesh within the stem is firm and creamy white.

SIZE: Cap is 5–10(20) cm broad; stem is 6–10(15) cm long and (1)2–3(4) cm thick; pores are 2-4 per mm.

LOOK-ALIKES: The edible but uncommon *Albatrellus confluens* has a similar form but differs in that the cap is pale orange rather than blue/green.

OCCURRENCE: In the late fall this mushroom appears either alone or as fused clusters on the ground, reappearing in the same location year after year. It appears in mixed or conifer woods, where it grows in association with conifers.

COMMENTS: This mushroom has a mealy, starchy flavor and odor. It should be thoroughly cooked before eating in order to soften the otherwise tough flesh.

Admirable Bolete

syn. Velvet Top, Bragger's/Admiral Bolete

- cap is velvety/rough
- underside has large yellow pores
- stem is darkly streaked and bark-like
- pores and flesh do not discolor

DESCRIPTION: The short-lived "admiral" bolete commands respect, both for its stature and flavor. The fleshy **CAP** is initially convex with inrolled edges. In maturity it becomes flat with the edges often retaining a flap of tissue from the partial veil. The cap is initially dark reddish brown or dark chocolate brown. In age the colors become more faded and the cap may develop pinkish blotches. The surface texture is dry and rough/velvety, hence the common name "velvet top." In older specimens, the cap can develop patches resembling dry-cracked mud. The firm but watery **FLESH** within the cap is white to yellowish. When cut it either turns pinkish or does not discolor.

The **UNDERSIDE** of the cap has a spongy pore surface that is depressed around the stem. The pores constitute the open ends of a tube layer. They are pale yellow when young, then olive-yellow in maturity and do not discolor significantly when bruised. The **SPORE** deposit is olive brown. The **STEM** is solid (not hollow) and tapers towards the top. It is similar in color to the cap except at the base, where it is more yellowish. The surface of the stem is darkly streaked with brown ridges/grooves that run longitudinally. It's texture is rough and pitted, and near the top there is often a network of raised ridges that create a ragged, net-like pattern, called a "reticulum." The flesh within the stem is pinkish to yellowish and does not discolor when cut.

SIZE: Cap is 5–15(20) cm broad; pores 1–2 mm wide; stem 7–15(20) cm long, 1–3(5) cm thick.

LOOK-ALIKES: The inedibly bitter *Boletus coniferarum* differs in that the flesh and pores bruise blue, the stem is white at least on the upper portion, and the pores are smaller (2/mm). The probably edible *Boletus fibrillosus* differs in that the top of the stem is yellowish, the bottom is whitish, and the base has a pinched terminus.

OCCURRENCE: This species has a short-lived fruiting period that occurs in the late summer or fall, soon after rainy periods. It appears either alone, scattered, or in small groups and typically grows in mossy areas, often on or near the decaying stumps and logs of conifers, particularly **western** and **mountain hemlock**.

COMMENTS: Younger specimens are best for collecting, since older ones quickly become infested with fly larvae. This species is also attacked by the powdery white/yellow mold *Hypomyces chrysospermus*, rendering it inedible. When cooked this mushroom has a juicy consistency and a pleasant lemony flavor that is strongest in the skin. The tube layer is usually scraped away before cooking.

Zeller's Bolete

syn. Zeller's Boletus

- cap is dark brown to nearly black
- young cap is velvety with a greyish bloom
- underside has large yellow pores
- stem is red with whitish base

DESCRIPTION: The **CAP** is initially convex and becomes broadly convex to nearly flat in maturity. When young it is dark brown to nearly black and powdered with a greyish bloom. In maturity it often develops reddish hues, especially towards the margin or wherever cracks appear. The surface texture is initially velvety and wrinkled, however the wrinkles smooth away in maturity. The **FLESH** within the cap is white to pale yellow. It slowly and inconsistently discolors to blue when cut.

The **UNDERSIDE** of the cap has spongy pores that constitute the openings of a tube layer. The pores are angular and relatively large. Their color ranges from yellow to olive yellow and they usually discolor to blue when bruised. The **SPORE** deposit is olive-brown. The fibrous **STEM** is solid (not hollow) and non-tapering. It is initially yellow, then becomes overlaid with a reddish flush or reddish, longitudinal streaks that eventually turn the entire stem red except for a white or yellow zone at the base. Unlike some boletes, it does not have a net-like pattern of ridges (reticulum) at the top of the stem. The flesh within the stem has the same color as the stem surface. It slowly and inconsistently discolors to blue when cut.

SIZE: Cap is 5–10(15) cm broad; pores are 1–2 mm wide; stem is 5–8(10) cm long and 7–13(25) mm thick.

LOOK-ALIKES: The edible and very similar *Xerocomellus atropurpureus* differs in that the cap has tones of dark purple and is not white at the edges. The edible *Xerocomellus chrysenteron* differs in that the cap is dark brown to olive gray and soon develops shallow, mosaic-like cracks, like dried mud, with redness showing through the cracks. The edible *Boletus truncatus* differs in that the stem is yellowish at the top, purplish red at the base and the cap is dark olive to olive-brown. When old, the cap exhibits cracks and may develop reddish tones near the edges.

OCCURRENCE: In the late summer and early fall, this common bolete appears either alone, scattered, or growing in small groups. It grows in mixed conifer forests, particularly in association with **Douglas-fir**, but is also sometimes found under hardwoods.

COMMENTS: This bolete has a flavor that is somewhat lemony. Because the texture is slightly mucilaginous, it may not be to everyone's taste or may be preferred after drying. When dried, it has a strong but pleasant lemony/buttery odor. Only younger specimens should be collected, since older ones tend to be wormy. This mushroom is a good candidate for bulking up a mixed mushroom dish. As with other boletes, the tube layer is usually scraped away before cooking.

King Bolete

syn. Porcino, Porcini, Cep, Penny Bun

- cap is smooth/slippery, has white rim
- underside of cap has tiny, round pores
- pores are white, then yellowish in age
- pores and flesh do not bruise blue
- stem has white, net-like pattern

DESCRIPTION: Ranked as a gourmet mushroom, the "king" is a wonderful find. The **CAP** is initially convex and becomes broadly convex to nearly flat in maturity. Its color ranges from tan to reddish brown, typically darkening as it matures but often retaining a whitish rim. The surface is smooth and slippery to the touch – more so in moist conditions. When old it can become uneven and irregularly pitted. The **FLESH** within the cap is white and does not discolor when cut.

The **UNDERSIDE** of the cap, when examined closely, reveals tiny, round pores, which are the openings of the tube layer. The underside is initially white and "stuffed" (covered with a cottony white coating). As it ages, the pore surface slowly becomes yellow-olive and then olive brown. When bruised, it discolors to brownish, if at all. The **SPORE** deposit is olive brown. The **STEM** is solid (not hollow) and varies in shape. In some forms it is greatly swollen, however it may be equally thick throughout its length or enlarged and bulbous. The stem is white to pale brown (but not yellowish) and at least on the upper third it bears a characteristic fine, white, net-like pattern, called a "reticulum." No veil or ring is present.

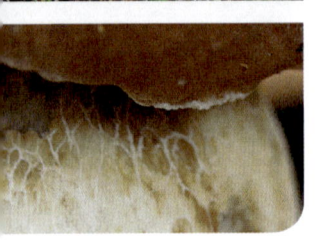

SIZE: Cap is 5–20(30) cm broad; pores are 2-3 per mm; stem is 8–17(25) cm long and up to 7 cm thick at the base.

LOOK-ALIKES: A number of inedible boletes are superficially similar in appearance but lack one or more of the key characteristics. The probably edible *Boletus fibrillosus* differs in that the cap is dark brown, velvety/fibrous, the stem is brown except at the yellow apex, and the pores on the underside are yellow at all stages. The edible *Boletus regineus* differs in having an almost black, burnt-looking cap. See also *Boletus rex-veris* (p. 11).

OCCURRENCE: In the summer and fall this mushroom appears either alone, scattered, or in small clusters. It often grows in association with conifers and is widely distributed.

COMMENTS: This sought-after mushroom has a nutty/meaty flavor, smooth texture and no distinctive odor. It is highly perishable due to its high water content and becomes slimy and soft when older. Scrape away the tube layer but do not peel or wash before cooking. These versatile mushrooms can be preserved by drying, freezing, pickling, or canning.

Spring King Bolete

- grows in mountains in the spring
- cap is dry and rusty brown
- pores on underside are yellow, bruise rusty
- top of stem has net-like pattern
- flesh does not change color when cut

DESCRIPTION: The **CAP** is initially bun-shaped with edges that are incurved. In maturity the cap becomes broadly convex, then nearly flat in age. The cap ranges in color from rusty brown to wine-brown. It is often darkest near the center, though areas covered with duff or soil may remain pale. Young caps are often coated with a very fine, white bloom. The mature cap has a dry texture and may be irregularly wrinkled and/or pitted. The **FLESH** within the cap is firm and white. It does not change color when cut.

The **UNDERSIDE** of the cap has a sponge-like surface of pores, which are the openings of a tube layer. The pore surface is depressed where the stem attaches. The pores are circular to angular. They are initially white and "stuffed" (covered with a cottony white coating). Over time they become yellow, then olive-yellow, and may become cinnamon brown near the cap edge. When bruised, the pores discolor to rusty brown. The **SPORE** deposit is dark olive to olive brown. The **STEM** is solid (not hollow), may be enlarged downwards, and is typically pointed at the base. The stem is white when young. Over time it becomes pale yellow near the top and pale brown below. The upper part of the stem exhibits a fine, fishnet-like pattern of reticulation on the surface. When handled or bruised the stem becomes brownish. No veil or ring is present.

SIZE: Cap is 10–30 cm wide; pores are 2-3 per mm; stem is 5–20 cm long and 2–10 cm thick.

LOOK-ALIKES: A number of inedible boletes are superficially similar in appearance but lack one or more of the key characteristics. The probably edible *Boletus fibrillosus* differs in that the cap is dark brown and it grows in coastal forests. The edible *Butyriboletus regius* differs in that the cap is pink to rose red, the stem is bright yellow, the pores slowly and inconsistently discolor to blue when bruised, and it occurs in coastal forests. See also *Boletus edulis (p. 10)*.

OCCURRENCE: This fairly common mushroom appears either alone, scattered, or in clusters on the ground in the spring and early summer. It grows in association with mountain conifers, especially **ponderosa pine**. It can often be found beneath understory shrubs and buried in needle duff. Look for it on the east side of the Cascades.

COMMENTS: This mushroom is considered a choice edible and is similar in flavor to other members of the *Boletus edulis* complex. The odor is not distinctive.

White King Bolete

- cap is suede-like, lacks a skin layer
- top of stem has white, net-like pattern
- underside of cap has small, white pores
- pores do not discolor when bruised
- flesh does not discolor when cut

DESCRIPTION: The **CAP** is initially convex with edges that are incurved. As the cap expands, it becomes broadly convex, then flattens in age. The cap color is dull white. When bruised, it discolors to brownish, if at all. In maturity the cap surface is smooth to slightly velvety or has a suede-like texture. The flesh within the cap is white and does not discolor when cut.

The **UNDERSIDE** of the cap, when closely examined, reveals a spongy surface that is composed of small, angular pores. The pores are the openings of a tube layer which is the site of spore production and dispersal. When young, the tube layer is "stuffed" (covered with a white coating). As they mature, the pores change in color from white to yellowish, then become olive/brownish. They do not discolor when bruised. The **SPORE** deposit is olive brown.

The **STEM** is solid (not hollow) and has a variable shape. When young it either widens downwards or is bulbous at the base. In maturity it becomes more or less equally thick throughout its length. The stem is white at the base; elsewhere, it is either whitish or tinged pink/cinnamon. The stem exhibits a fine net-like pattern (reticulation) at least on the upper part. The reticulation is white to pale brown. No veil or ring is present.

SIZE: Cap is 6–20(30) cm wide; stem is 8–15 cm long and 2–5 cm thick at apex.

LOOK-ALIKES: Whitish specimens of the edible *Boletus edulis (p. 10)* can have a similar appearance, but differ in that the cap is shinier and has a skin layer. The edible *Leccinum holopus* differs in that the cap is smooth to tacky, and the stem exhibits dark, projecting scales called scabers, comparable to the stem of *Leccinum insigne (p. 16)* or *Leccinum scabrum (p. 17)*.

OCCURRENCE: In the spring, fall, and winter this fairly common mushroom appears either alone, scattered, or in groups on the ground. It can be found growing in association with **spruce** and live **oak**.

COMMENTS: This mushroom is a choice edible. It has a pleasant, nutty flavor and mild to strong odor that can become unpleasant when it is dried.

Fat Jack

syn. Blue Staining Suillus

- underside of cap has large, yellow pores
- pores become reddish-brown on bruising
- stem is yellow above ring, brownish below
- flesh in stem turns bluish green when cut

DESCRIPTION: The **CAP** is initially convex. In maturity it becomes broadly convex to flat with edges that sometimes retain hanging remnants of the veil. The cap color is dull orange-brown or yellow-brown and is more yellowish closer to the edge. The surface of the cap is slippery when wet, and is often streaked. The **FLESH** within the cap is pale yellow. When cut, it either does not change color or flushes pinkish.

The **UNDERSIDE** of the cap does not have gills but instead has a spongy surface of pores that may extend slightly onto the stem. The pores constitute the openings of a tube layer, which is the site of spore production and dispersal. The pores are large, irregularly angular, and are often radially arranged. They are yellow to honey colored and darken as they age, becoming dingy brown when old. When bruised, the pores slowly discolor to reddish brown. The **SPORE** deposit is cinnamon brown. When young, the pores are covered by a whitish partial veil. The veil leaves an uneven, band-like ring on the stem. The ring is initially white and soon becomes the same color as the cap. The **STEM** is solid (not hollow) and may taper in either direction. Above the ring, the stem is the same color as the pores and is often weakly reticulate near the top (i.e., exhibits a net-like pattern). Below the ring the stem is duller in color and matted with appressed fibers, making it dingy brown and mottled with reddish or brown stains. When bruised, the stem discolors to brown. The flesh within the stem is yellow. It slowly discolors to bluish green when cut, particularly near the base.

SIZE: Cap is 6–14 cm broad; pores are 1½–2 mm long and ~1 mm wide; stem is 2½–8 cm long and 2–3 cm thick.

LOOK-ALIKES: The edible and similar-looking *Suillis ponderosus* differs in that the cap is smoother, the stem base is greenish when cut, and young specimens have a veil that is slimy or sticky. See also the edible *Suillus lakei (p. 15).*

OCCURRENCE: In the fall this common mushroom appears either alone, scattered, or in groups on the ground. It grows in mossy areas under mixed conifers, particularly where **Douglas-fir** is present.

COMMENTS: This mushroom has a flavor and odor that ranges from lemony to slightly sour. Those who find its texture to be unpleasantly slimy may prefer to dry it before using it in mushroom dishes.

Short-stemmed Slippery Jack

syn. Stubby Stem

- cap is slippery or sticky when moist
- cap edges are incurved and smooth
- underside of cap has pale yellow pores
- pores and flesh do not discolor when cut
- stem is white or yellowish and lacks ring

DESCRIPTION: The **CAP** is hemispheric when young, then becomes broadly convex with incurved edges that are free of any adhering veil remnants. In age, the cap becomes almost flat. The cap color is initially dark chocolate brown and becomes a lighter brown as it ages. The skin layer is shiny and peelable. In moist conditions it is covered with a sticky slime layer. The **FLESH** within the cap is white to yellowish and does not discolor when cut.

The **UNDERSIDE** of the cap, when closely examined, reveals a spongy surface of pores, which are the openings of the tube layer. The pores are small, round, and initially pale yellow, then dingy olive in maturity. They do not discolor when bruised. The **SPORE** deposit is dull cinnamon. In young specimens the **STEM** is short and thick (hence the name "short-stemmed"), however in mature specimens the length is more variable. Even so, it's usually shorter than the diameter of the cap. The stem is all-white when young and becomes yellowish as it ages. The flesh within the stem does not discolor when cut.

SIZE: Cap is 5–10 cm broad; pores are 1–2 per mm; stem is 2–5 cm long and 1–3 cm thick.

LOOK-ALIKES: The probably edible *Suillus pseudobrevipes* differs in that the cap is honey-colored, the cap edge retains remnants of the partial veil, and a ring is present on the stem. The edible *Suillus albivelatus* differs in that the cap is orangey brown, the stem has a fleeting ring zone, and the cap edge retains remnants of the veil. The edible *Suillus granulatus* differs in that the cap is more broadly convex, the cap's coloration is less even, and the stem has prominent glandular dots when young. The edible *Suillus borealis* differs in that the edge of the cap retains remnants of the partial veil, at least for some time, and the stem discolors to ocher or reddish brown when bruised/handled.

OCCURRENCE: In the late summer and fall this species appears either alone, scattered, or growing together in clumps. It prefers sandy soils and grows in association with various species of two- and three- needle pine. It is one of the most common species of *Suillus*.

COMMENTS: This bolete has a pleasant flavor that is mild but somewhat lacking in texture. No species of *Suillus* are known to be poisonous. However, some people have experienced gastrointestinal upset upon consumption of the slime layer on the skin. For that reason, remove the skin before cooking. The tube layer is also usually scraped away before cooking, since it does not have a fleshy texture; on smaller/younger mushrooms the tubes can remain.

Western Painted Suillus

syn. Matte Jack, Lake's Bolete

- cap is dry and scaly
- pores are large and bruise brown
- stem has a thin ring
- cap's flesh does not discolor when cut

DESCRIPTION: The fruiting body has a fleshy **CAP** that is initially convex with inrolled edges. As it matures it becomes broadly convex, then flat with a shallow depression in the center. The edge of the cap retains hanging remnants of the veil and sometimes becomes upturned in age. The cap surface is covered with rusty brown, appressed scales, giving it a scruffy appearance. The surface is dry, which is unusual for a *Suillus*; however, if the scales weather away, a smooth, gelatinous sublayer is exposed. The **FLESH** within the cap is pale yellow. It does not discolor when cut, or weakly discolors to pinkish.

The **UNDERSIDE** of the cap has spongy pores that are the openings of a tube layer. The pores are large, angular and radially elongated. The pore surface is dingy ocher, bruising brownish. The tubes, which are relatively shallow, may extend slightly onto the stem. The **SPORE** deposit is cinnamon brown. When young, the pores are covered by a delicate, whitish, partial veil. It soon disappears, leaving a thin and transient whitish ring on the upper part of the stem. The **STEM** is solid (not hollow) and is equally thick throughout its length. Above and below the ring it is yellow, soon becoming brown from handling. The flesh within the stem is yellowish. In young specimens it inconsistently and weakly discolors blue/green when cut.

SIZE: Cap is 6–15(20) cm broad; pores are 1–2 mm wide; stem is up to 3–8(12) cm tall and 1–3(4) cm thick.

LOOK-ALIKES: The probably edible *Suillus ponderosus* differs in having a cap that is smooth and sticky, and a veil that is sticky and orange/yellow. See also *Suillus caerulescens (p.13)*.

OCCURRENCE: In the fall this common mushroom appears either alone, scattered, or in small groups. It grows in poor, exposed soil and grassy areas in association with **Douglas-fir.** It is one of the most common species of *Suillus*.

COMMENTS: Although it has its enthusiasts, this species is usually considered a mediocre edible. Those who object to its slimy texture may prefer to dry it before using it in mushroom dishes. Because these mushrooms are soon infested, seek out younger specimens. Before cooking, peel away the skin and scrape off the tube layer.

Aspen Scaber Stalk

syn. Aspen Bolete

- cap is a variable shade of orangey brown
- stem is marked with brown/black scabers
- cap-underside is whitish, then yellow/brown
- cut flesh discolors purplish, then blackish

DESCRIPTION: The **CAP** is initially convex and has a rim of tissue around the edge. As it matures it becomes broadly convex. The color is a variable shade of orangey brown. The cap surface is initially smooth or finely velvety, and either remains so or becomes more roughly textured in maturity. When old it becomes very finely cracked, liked dried mud. The **FLESH** within the cap is white. When cut it slowly and erratically discolors to purplish grey, then black without an intervening reddish phase.

The **UNDERSIDE** of the cap has small, sponge-like pores. The pores are the openings of a tube layer, which is the spore-bearing tissue. The pores are round and initially whitish, then slowly become dingy yellow and finally yellow-brown to olive-brown in age. When bruised, they either remain the same color or become yellowish or brown. The **SPORE** deposit is dark yellow-brown to olive-brown. The tube layer is narrowly to deeply depressed around where the stem attaches, more so in age. The **STEM** is solid (not hollow), often widen downwards, and is often swollen at the base. The stem is white and is marked with protruding scales called "scabers." The presence of scabers is a characteristic of species of *Leccinum*. The scabers are lightly colored when young, then become dark brown to black. The base of the stem often discolors to blue when bruised. The flesh of the stem is rather tough and fibrous. No partial veil or ring is present.

SIZE: Cap is 4–15 cm broad; stem is 7–12 cm long, 1–2(3) cm thick at top; pores are 1–3 per mm.

LOOK-ALIKES: A number of orange-capped species of *Leccinum* are similar in appearance, none of which are known to be poisonous (however see caveats regarding edibility in comments). The very similar and edible *Leccinum aurantiacum* differs in that the cut flesh discolors to reddish or pink before turning black. The edible *Leccinum ponderosum* differs in that the cap is more rusty red and the cap flesh does not discolor when cut. The edible *Leccinum manzanitae* differs in that the cap is reddish brown and sticky, and it grows with **manzanita** and **madrone**.

OCCURRENCE: In the spring, summer and fall this common mushroom appears either scattered or in groups on the ground. It grows in association with **aspen** and to a lesser degree **birch**.

COMMENTS: This mushroom has a non-distinctive odor and flavor. Although it is considered edible, orange species of *Leccinum* have been implicated in random cases of severe gastrointestinal upset, either due to undercooking or individual sensitivity. Be sure to thoroughly cook this mushroom (especially the stem) before eating, and begin by eating only a small amount. Like most boletes, this species is soon infested, so younger specimens are best for eating.

Birch Bolete

syn. Scaber Stalk, Rough-stemmed Bolete

- cap is a variable shade of dull brown
- stem is marked with brown/black scabers
- pores under cap are whitish to yellowish
- flesh and pores don't change color when cut
- grows under birch

DESCRIPTION: The **CAP** is initially convex, then becomes broadly convex with edges that may overhang slightly but lack any hanging veil remnants. The cap is dull grey brown to dull yellowish brown and often develops olive tones when old. The surface is suede-like when young, smooth in maturity, and sticky when moist. In age it may become cracked. The **FLESH** within the cap is white when young and dingy brownish in age. When cut, it either does not discolor or slowly discolors to dingy brownish.

The **UNDERSIDE** of the cap has a spongy, porous surface that is deeply depressed around where the stem attaches. The pores are the openings of a tube layer that is the spore-bearing surface. The pores are small, round or irregular in shape. When young they are whitish, then become greyish to olive/brown over time. When bruised, the pores either do not discolor or slowly discolor to brownish. The **SPORE** deposit is olive/brown. The **STEM** is relatively long, solid (not hollow) and widens slightly towards the base. The surface is roughened by dark brown to blackish "scabers" over a white base color. The scabers are fine near the top of the stem and become rougher, coarser and darker in the lower part. The flesh within the stem is white. When cut the upper part does not discolor significantly.

SIZE: Cap is 4–10 cm broad; stem is 7–12(14) cm long, 7–12(16) mm thick; pores are 2–3/mm.

LOOK-ALIKES: The presence of scabers on the stalk and pores on the underside of the cap are key characteristics of species of *Leccinum*, none of which are known to be poisonous (however see caveats regarding edibility in comments). The edible *Leccinum manzanitae* differs in that the cap is sticky and reddish brown and it grows in association with **manzanita** and **madrone**. The probably edible *Leccinum fibrillosum* differs in that the flesh of the cap stains pink, then darkens to dull purplish when cut.

OCCURRENCE: In the summer and fall this common mushroom appears either alone, scattered, or in groups on the ground. It grows under hardwoods, and particularly birch. It can be found in urban/suburban environments and may be abundant when other mushrooms are not.

COMMENTS: This mushroom is considered edible, however it has been implicated in random cases of severe gastrointestinal distress, either as a result of undercooking or individual sensitivity. A mediocre edible, it has a mild taste and an indistinct odor. Firm young specimens are preferred for eating. Be sure to thoroughly cook this mushroom (especially the stem) and begin by eating only a small amount.

Saffron Milk Cap

syn. Red Pine Mushroom

- cap has concentric orange zones
- flesh exudes orange milk when cut
- flesh and gills discolor to dark red, then eventually green when cut
- spore deposit is creamy yellow

DESCRIPTION: The **CAP** is initially convex with inrolled edges and a depressed center. In maturity it becomes broadly convex with undulating edges and usually retains the central depression. When old it becomes funnel-shaped with uplifted edges. The cap ranges in color from orange to brownish orange and typically exhibits mottled, concentric color zones as well as patchy greenish stains where bruised. The surface is smooth and sticky when moist. The **FLESH** within the cap is firm, grainy/crumbly, and ranges in color from orange to yellowish. When cut, the flesh characteristically ooze a bright carrot orange milk, or "latex." The cut flesh discolors to wine red, then slowly to greenish over the course of an hour or so.

The **UNDERSIDE** of the cap has fragile, crowded gills that are broadly attached to the stem and may descend slightly down onto the stem. The gills are orange. When bruised they discolor to wine red, then slowly to green. The **SPORE** deposit is creamy yellow. The **STEM** is roughly equal in thickness throughout its length. It is initially stuffed with loose tissue, then becomes hollow in age. The color of the stem is similar to that of the cap but may be paler in the upper part. The stem surface is smooth and sometimes sparingly marked with darker pits, depressions and/or green stains. When bruised, it discolor to wine red, then slowly to green. The base of the stem has appressed white tissue. The flesh within the stem is pale orange.

SIZE: Cap is 5–14(35) cm broad; stem is 2–7 cm long and 1–3 cm thick.

LOOK-ALIKES: A number of poisonous and/or acrid species of *Lactarius* are similar in appearance but differ in that the cut flesh exudes a white rather than orange milk and the flesh does not eventually discolor to greenish. The edible *L. rubrilacteus* differs in that the cap and latex are more reddish. The edible and similar-looking *L. deterrimus* differs in that the cut flesh exudes an orange milk that stains the surface purplish, then dull, dark green; it is sometimes considered a variety of *L. deliciosus.*

OCCURRENCE: In the late summer and fall, this mushroom appears either alone, scattered, or in small clusters on the ground. It grows in the duff in mixed conifer woods, especially with pine. It is widespread in montane to subalpine regions.

COMMENTS: This mushroom has a slightly fruity odor and a grainy or soapy texture. The flavor is bland to slightly bitter, but not unpleasant. Young mushrooms are preferred since older specimens can be maggot-ridden. Slow cook for best results. Be warned that eating this mushroom may turn your urine red.

Lobster Mushroom

- surface is red, rough and pimpled
- shape is often contorted
- gills are reduced or absent

DESCRIPTION: This species is a mold-like parasite that selectively attacks certain mushrooms, rapidly transforming them into red "lobster" mushrooms. A lobster mushroom typically becomes twisted away from its original shape, sometimes into the shape of an inverted pyramid. The entire **SURFACE** is bright orange-red, reminiscent of a cooked lobster shell. Sometimes the surface of the cap is more orangey than the underside and stem, which are more deeply reddish. The exterior is firm to the touch, coarse, and roughened with tiny pimples. Over time it may also become cracked. The **FLESH** within the cap and stem ranges from white to orange-tinted. If the host mushroom is a species of *Lactarius* it may exudes a white, milky liquid. If the host is a species of *Russula*, the flesh is often denser and less crumbly than it was originally.

The **UNDERSIDE** of the lobster mushroom has gills that are typically reduced to blunt folds or disappear entirely. The **STEM** may also be reduced or disappear. The **SPORE** deposit, which is difficult to collect, is creamy white.

SIZE: Fruiting body is 5–15 cm broad.

LOOK-ALIKES: None.

OCCURRENCE: This common mushroom appears in the summer or fall shortly after rainy weather and is often scattered in significant numbers. The parasite is only known to target edible species of *Lactarius* and *Russula* mushrooms, particularly *Russula brevipes*. Lobster mushrooms can be found in hardwood and coniferous forests throughout the Pacific Northwest.

COMMENTS: Although considered edible, the lobster mushroom can cause gastric upset in some individuals. It has a distinctively fishy or shellfish-like aroma and a flavor that is highly dependent on the host mushroom. After collecting, remove any dirt that might have become lodged in cracks of the cap, which could lead to rot. Be sure to discard specimens that are soggy rather than firm and crisp, or any that smell foul, which indicates decay. This mushroom is best preserved by cooking and then freezing rather than dehydrating. When cooked it imparts an interesting aroma and color, and adds a crunchy texture to dishes.

Green Brittlegill

syn. Grass Green Russula, Tacky Green Russula

- flesh and gills are brittle/crumbly
- skin of cap peels halfway to center
- spore deposit is creamy/yellowish

DESCRIPTION: The **CAP** is initially convex. As it matures it flattens, becomes cushion-shaped, and develops striated edges. In age the central depression deepens and the cap becomes almost funnel-shaped, with the edges often splitting. The color is characteristically a shade of green, ranging from grey-olive to grass green. Near the edges, the color is sometimes suffused or spotted with brownish or yellow. The skin of the cap is smooth, may be slightly velvety at the center, and becomes tacky in wet weather. It can be peeled one third to halfway from the edge to the center. The **FLESH** within the cap is firm, white, and brittle/crumbly, which is a characteristic of species of *Russula*. The flesh may be tinged the same color as the cap close to the surface.

The **UNDERSIDE** of the cap has crowded gills that are forked near the stem. The gills are variable in how they attach; they can either be free or almost-free from the stem, broadly attached, or be notched at the attachment point. The gills are initially white. In age they become tinged with pale yellow and/or become spotted or brown-stained. The **SPORE** deposit is creamy to pale yellow. The **STEM** is equally wide throughout its length or may taper slightly in either direction. It ranges in color from white to slightly greenish or yellowish. The stem discolors to pale olive brown when handled or bruised. At the base it sometimes develops reddish spots. The stem surface is smooth with very fine longitudinal wrinkles. The stem is initially stuffed with loose fibers, and becomes hollow in age.

SIZE: Cap is 3–10 cm broad; stem is 4–6 cm long and 1–2 cm thick.

LOOK-ALIKES: A number of other edible species of *Russula* are similar in appearance. The edible *Russula heterophylla* differs in that the spore deposit is white and it grows with **beech**, **oak**, or **chestnut**. The edible *Russula cyanoxantha* differs in that the cap often has purplish tones, the spore deposit is white, and it usually grows under **beech**. The edible *Russula olivacea* differs in that the cap is larger (8–16 cm broad), the cap is olive green (with reddish, brownish or purplish tones when older), and the spore deposit is deep yellow to orange. **TIP:** Species of *Russula* can be differentiated from most non-*Russula* look-alikes by confirming that the stem is brittle and snaps cleanly in two, like a piece of chalk, rather than bending or fraying lengthwise.

OCCURRENCE: In the summer and fall this mushroom can be found growing either alone, scattered, or in groups on the ground. It can be found in the humus or moss carpets of the forest floor in coniferous and deciduous woods.

COMMENTS: This mushroom has a flavor that is slightly peppery, more so when young. It has a mild, indistinct odor and should be well cooked, since it is mildly poisonous when raw.

Aniseed Toadstool

syn. Aniseed Funnel, Blue-green Anise Mushroom

- cap and gills are blue/green
- smells strongly of anise (like licorice/ouzo)
- spore deposit is pinkish
- warning: contains small amounts of the toxin muscarine.

DESCRIPTION: The mushroom has a **CAP** that is initially convex with edges that are incurved. As it matures, the cap expands to become flat, then becomes shallowly depressed in age with edges that are broadly wavy. The cap color is bluish green to greenish. It fades quickly as it ages or in dry weather. The surface of the cap is smooth but not sticky. The **FLESH** within the cap is whitish or tinged the same color as the cap.

The **UNDERSIDE** of the cap has gills that are broadly attached and may extend slightly down onto the stem. The gills are fairly closely spaced, and occasionally forked. They are typically colored like the cap but may also be whitish, pinkish or pale buff. The **SPORE** deposit is pinkish cream. The **STEM** is sometimes bent, or flattened. It may or may not widen downward to a somewhat enlarged base. The surface of the stem is whitish and may be striated with fibers. The stem base is spongy, often enlarged, and covered with white, downy mycelium. The stem is stuffed with loose tissue when young and becomes hollow in age. No veil or ring is present.

SIZE: Cap is 2–11 cm wide; stem is 2–6(9) cm long and 5–15(30) cm thick.

LOOK-ALIKES: A number of greenish mushrooms, some of which are inedible or poisonous, are superficially similar in appearance but lack the strong and distinctive aniseed odor. For example, the poisonous/suspect *Stropharia aeruginosa* differs in that it the cap is slimy, the spore deposit is brown, the stem has a dark ring, and no aniseed odor is present. The inedible/suspect *Stropharia pseudocyanea* differs in that the caps are slimy, the gills are notched, the stem has white scales below a ring zone, and the spore deposit is purple-black. The inedible *Stropharia caerulea* differs in that the cap is more rounded (convex to conical), the spore print is brown, the stem has white scales below a ring zone, and no aniseed odor is present.

OCCURRENCE: In the summer and fall this mushroom appears in the soil, either alone, scattered, or in small clusters where the stems are not fused. It can be found in the debris under both conifers and hardwoods.

COMMENTS: When fresh, this mushroom has a fragrant odor and taste of aniseed. Because of its strong taste and the fact that it contains small amounts of the toxin muscarine, it is best used as a flavoring agent rather than eating it in quantity.

Inky Cap

syn. Ink Cap, Alcohol Inky Cap,
Tippler's Bane

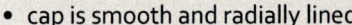

- cap is smooth and radially lined
- cap becomes black and inky when old
- clusters at base of trees/stumps
- spore deposit is blackish
- warning: do not consume with alcohol

DESCRIPTION: The **CAP** is initially oval or egg-shaped. In maturity it becomes broadly conical to bell-shaped with edges that are often tattered or split, but free of any clinging tissue. The color of the cap ranges from cream to lead grey or brownish, with the apex sometimes being darker. When old, the edges of the cap may become upturned as it dissolves into a black, inky mass from the edges inwards. The surface has a smooth to silky texture and faint lines or grooves that radiate to the edges. The **FLESH** within the cap is thin, soft, and greyish.

The **UNDERSIDE** of the cap has very crowded gills that are nearly free from the stem. The gills are initially whitish, then lavender-grey, and finally black as they dissolve away. The **SPORE** deposit is blackish. In young specimens a white partial veil covers the gills. The veil leaves a transient ring near the base of the stem. The **STEM** is smooth, hollow, and equally wide throughout its length. It is whitish near the top and grey or brownish at the base.

SIZE: Cap is 3–8 cm broad; stem is 6–15 cm long, ¼–1 cm thick.

LOOK-ALIKES: The edible *Coprinellus disseminatus* differs in that it is smaller (under 2 cm broad) and does not become inky when old. The less common *Coprinopsis romagnesiana* is similar in that it clusters at the base of trees, also has blackish spore print, and also has a cap that becomes inky. However, it differs in that the cap has a tan ground color as is covered with brown, appressed scales.In terms of edibility and interactions with alcohol, it is similar to this species. See also *Coprinellus micaceus (p. 23)*.

HABITAT: In the spring, summer and fall this common mushroom appears after rainy periods. It grows in tight clusters from stumps, decayed wood or buried wood, such as from the roots of deciduous trees. It is often found in disturbed soil in urban areas, including near gardens, footpaths and other grassy areas where wood has become buried.

COMMENTS: This species is often not recommended for eating since it contains the compound coprine, which causes nausea, palpitations and flushing when alcohol is consumed either before or afterwards. It was once administered as a treatment to alcoholics to dissuade them from their habit. Both the flavor and odor are mild. When pan-fried, it dissolves into a black, pulpy mess. Do not consume this mushroom unless you are abstaining from alcohol.

Glistening Ink Cap

syn. Glistening Inky Cap,
Mica Cap, Shiny Cap

- cap has radial grooves
- young cap is covered in granules
- cap becomes black and inky when old
- clusters at base of trees/stumps
- spore deposit is blackish

DESCRIPTION: When young the **CAP** is oval or egg-shaped. In maturity it becomes bell-shaped to convex with edges that are often split or tattered. The cap color is honey brown or amber with grey tones at the edges and russet/brown tones in the center; in moist conditions it darkens overall. When old, the edges sometimes become upturned as the cap blackens from the edges inwards. The upper surface is smooth and has conspicuous furrow-lines running from the edge to more than halfway to the center. In younger specimens the surface is characteristically powdered with glistening, salt-like granules, which soon wash or fall away. The **FLESH** within the cap is thin, soft, and whitish.

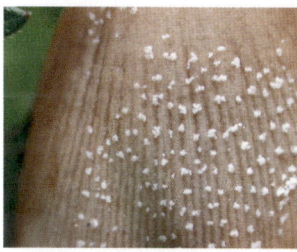

The **UNDERSIDE** of the cap has crowded gills that are either attached to the stem or nearly free from it. The gills are initially whitish, then purple-brown and finally black as they dissolve either partly or completely. The **SPORE** deposit is blackish. The fragile **STEM** is smooth, hollow, and equally wide throughout its length. It is silky white except at the base where it is buff.

SIZE: Cap is 2–5 cm; stem is 2½–8 cm long and 2–5 mm thick.

LOOK-ALIKES: A number of potentially poisonous mushrooms are superficially similar but do not match all the key characteristics. The less common *Coprinopsis romagnesiana* differs in that the cap is covered with darker brown appressed scales and lacks granules

when young. It interacts with alcohol in a similar way to *Coprinopsis atramentaria* (p. 22). *Coprinellus domesticus*, of unknown edibility, differs in that the young cap is dotted with whitish scales and the fungus forms a mat of orange, hair-like fibers on the decaying log that it grows from. The less common *Coprinellus flocculosus*, of unknown edibility, differs in that the young cap is covered with tufts of whitish, felty warts. The edible *Coprinellus disseminatus* is smaller (under 2 cm broad), lacks granules on the cap, and does not become inky when old.

HABITAT: In the spring, summer and fall this common mushroom emerges after rainy periods. It sometimes appears alone but more often forms large, dense clusters at the base of living or dying trees. Although it may appear terrestrial, it fruits from buried, decaying wood.

COMMENTS: The flavor and odor are mild and not distinctive. Because it lacks texture when cooked, this mushroom is not highly regarded as an edible. Collect specimens that have granules on the cap and/or are growing with older, inky specimens to reduce the chances of misidentification. Cook shortly after harvesting, before the mushroom begins to disintegrate.

Shaggy Mane

syn. Shaggy Ink Cap, Lawyer's Wig

- cap is tall and elongated
- cap has shaggy, upcurving scales
- cap is black and inky when old
- spore deposit is blackish

DESCRIPTION: The **CAP** is initially egg-shaped. In maturity it elongates and then becomes bell-shaped. It is creamy white in color and often has a brownish region around the apex. As it ages, the surface of the cap develops large, shaggy, up-curving scales. The edge of the cap becomes striated, tattered, and then upturned as it blackens and melts away towards the center, finally dissolving into an inky, black disc. The thin **FLESH** within the cap is soft and white.

The **UNDERSIDE** of the cap has tightly packed gills that are free from the stem. The gills are initially white, then pink and finally black. The **SPORE** deposit is blackish. In younger specimens a white, membranous veil covers the gills. It leaves a large, loose ring on the lower part of the stem. The ring becomes black from spores and often falls to the base of the stem. The **STEM** is straight, white, smooth and roughly equal in thickness throughout its length, terminating in a bulbous base. The stem is easily separated from the cap. The stem is either hollow or stuffed with hanging strands of loosely interwoven fiber.

SIZE: Cap is 3–6 cm broad and 5–10(15) cm tall; stem is 5–20 cm long and 1–2 cm thick.

LOOK-ALIKES: The similar-looking but not-recommended *Coprinopsis variegata* differs in that it grows on decaying hardwood and the cap usually has less of an overall white color. It may cause gastrointestinal distress and/or adverse reactions when consumed with alcohol, even in the space of several days. The inedible *Coprinopsis lagopus* has a similarly elongated cap but differs in that it is much smaller (under 4 cm broad when fully expanded). It has a very delicate fruiting body that lasts only a few hours before dissolving into black ink. A number of other poisonous or inedible white mushrooms that have warty or scaly caps are superficially similar. They differ in that they do not turn inky/black when old, and/or do not have a blackish spore deposit.

OCCURRENCE: This mushroom emerges in the early spring and late fall, shortly after rainy periods. It appears either alone, growing in lines, in densely packed groups, or forming fairy rings. It can be found in grassy urban areas such as lawns, fields, gardens, and by footpaths, but also occurs on bare ground, including on gravelly and hard-packed soil.

COMMENTS: This mushroom has a mild, delicious flavor, and works well when it is sautéed or breaded and fried. In rare cases it causes a reaction when consumed with alcohol. Young and fresh specimens should be used for cooking; they do not last long, even when refrigerated.

Shaggy Parasol

- cap has shaggy, brownish scales
- flesh turns orange/red when cut
- stem is hollow and has a double ring
- spore deposit is white

DESCRIPTION: The **CAP** is spherical when young. As it expands it nearly flattens but sometimes retains a central hump. Its color is grayish brown to olive brown, with a center that is grayish olive brown or dark reddish brown. The surface is smooth at the center and elsewhere has broad, sightly reflexed, concentric scales on a fibrous background, giving it a shaggy appearance. The edge of the cap soon breaks up and develops a torn appearance as a result of coarsely fibrous scales. The **FLESH** within the cap is whitish. When cut, it immediately discolors to orangey, then reddish.

The **UNDERSIDE** of the cap has broad gills that are free from the stem. The gills occur in two or three tiers and have edges that are finely fringed. The gills are white when young and become red to brown over time or when bruised. The **SPORE** deposit is white. The **STEM** is either equally wide along its length or widens towards an abruptly bulbous base. It is stuffed with loose fibers when young and over time becomes hollow. The mature stem has a thick, membranous, persistent double ring that is moveable on the stem. The edges of the ring are frayed or jagged. Above the ring, the surface of the stem is whitish and smooth or longitudinally fibrillose. Below the ring the stem is whitish and becomes spotted with reddish brown stains when bruised. When old, the stem turns gray to ocher brown. The flesh within the stem discolors to orangey red when cut.

SIZE: Cap is 5–12(18) cm across; stem is (7)9–15(18) cm long and 12–16(20) mm thick.

LOOK-ALIKES: The poisonous *Chlorophyllum molybdites* is similar in appearance but is rare to absent in Washington. It differs in that the gills are greenish when mature and the spore deposit is greenish. The edible *Chlorophyllum brunneum* differs in that the cap has brownish scales on a white background rather than on a brownish background and the stem exhibits a simple ring rather than a double-edged ring. The edible *Chlorophyllum rhacodes* differs in that the cap has brownish scales on a white background rather than a brownish background. See also *Coprinus comatus (p. 24)*.

OCCURRENCE: In the summer to fall this common mushroom appears either singly or in large numbers on the ground. It can be found growing in the litter of both coniferous hardwood forests as well as in parks, gardens, and in shrubberies.

COMMENTS: Although considered an excellent edible, this species may sicken some people. It has a mild, pleasant taste that is somewhat nutty. Its odor is reminiscent of raw potatoes.

Deceiver

syn. Waxy Laccaria

- cap, stem and flesh are dull orange
- gills are flesh-colored
- stem is tough, hollow and grooved/striate
- spore deposit is white

DESCRIPTION: This mushroom is common enough to be considered weed-like in some areas. It earns the common name "deceiver" for its highly variable form, color, and size. The **CAP** is initially convex with incurved edges. In maturity it either becomes flat or depressed in the center, where it sometimes develops a hole. The edges may become uplifted, wavy or striated. The cap is dull orangey brown, with darker tones sometimes occurring at the center. As it ages or dries out, it becomes paler overall, fading to buff. The surface of the cap is finely fibrous. The **FLESH** within the cap is thin and is similar in color to the cap surface.

The **UNDERSIDE** of the cap has gills that are usually well-spaced but sometimes are close together. They are broadly attached to the stem and may descend slightly onto the stem. The gills are flesh-colored and may have a faint purplish tinge when old. They are typically thick and somewhat waxy, but in the spirit of variability may also be thin; they can also be either narrow or broad. The **SPORE** deposit is white. The **STEM** is slender, tough, and more or less equally thick throughout its length. The stem is stuffed with loose fibers when young, later becoming hollow with thick, tough walls. The stem is the same color as the cap or is darker shade of reddish brown, and has downy, white tissue at the base. The stem surface is fibrous and often exhibits pronounced grooves or striations along its length. The flesh within the stem is similar to that of the cap. It does not change color when cut. No veil or ring is present.

SIZE: Cap 1–2½(6) cm broad; stem 2–6½(10) cm long, 3–10 mm thick.

LOOK-ALIKES: The inedible and less common *Laccaria montana*, *Laccaria pumila* and *Laccaria tortilis* differ in that they are smaller (under 35 mm broad) and prefer high elevations. The edible and similar-looking *Laccaria proxima* and *Laccaria nobilis* differ in that the cap is larger (up to 8 cm broad) and slightly scalier. The edible *Laccaria bicolor* differs in that the gills are faintly purplish. See also *Laccaria amethysteo-occidentalis* (p. 27).

OCCURRENCE: In the spring, summer and fall, this common mushroom appears either alone, scattered, or in tight or loose clusters. It grows with hardwoods and conifers, often in poor, sandy soils or in boggy areas.

COMMENTS: This mushroom has a taste that is mildly reminiscent of radish, and a similar odor. The stems are tougher than the caps and require extra cooking to soften. Some people discard the stems, though they constitute a good part of the bulk of the mushroom.

Amethyst Deceiver

syn. Western Amethyst Laccaria

- cap, gills, and stem are purple
- gills are well-spaced
- stem is tough and grooved/striate
- spore deposit is white

DESCRIPTION: The **CAP** is initially convex with an incurved margin, then becomes either broadly convex or nearly flat, sometimes with a depressed center. The edge of the cap may remain incurved in maturity. The color is a variable shade of purple that is influenced by the moisture level. When young it is deep purple, then fades to vinaceous and finally becomes buff or brown when old. The surface of the cap is finely fibrillose/scaly and may appear faintly lined when wet. The **FLESH** within the cap is similar to the surface color or is lighter grayish purple.

The **UNDERSIDE** of the cap has thick gills that are attached to the stem and are either sinuate or arched. The gills are well-spaced and variable in shape, ranging from narrow to broad. The gills are dark purple, fading to lilac in age and sometimes appearing waxy. The **SPORE** deposit is white. The **STEM** is roughly equal in thickness throughout its length. The surface of the stem is longitudinally striate or strongly grooved, and may be coarsely hairy or exhibit upcurved scales. The flesh within the stem is similar in color to the cap. No ring or veil is present.

SIZE: Cap is 1–7(9) cm broad; stem is 2–12 cm long and 5–12(15)mm thick.

LOOK-ALIKES: The common and poisonous *Mycena pura* can be similar-looking when it is young and still violet. It differs in that the cap is smaller, the stem is whitish, and it smells and tastes strongly of radishes. A number of other inedible/toxic species of *Cortinarius* and *Inocybe* are superficially similar but do not match all the key characteristics. See also *Laccaria laccata (p. 26).*

OCCURRENCE: In the fall and periods of colder weather, this mushroom appears either alone, scattered, or in numbers. It grows in association with with conifers, such as **Douglas-fir**, and is considered common in the Pacific Northwest in areas west of the Rocky Mountains.

COMMENTS: This mushroom has a mild flavor and odor that is non-distinctive to somewhat earthy. Some consider the stem to be too tough for eating, however when finely diced and thoroughly cooked they are suitable for eating.

Rosy Gomphidius

syn. Pink Gomphidius, Rosy Spike Cap

- cap is rosy pink and sticky
- gills extend onto stem
- stem base is yellow
- spore deposit is blackish

DESCRIPTION: This mushroom has a **CAP** that is initially convex with inrolled edges. In maturity it becomes broadly convex to flat or slightly depressed in the center with the edges of the cap sometimes becoming uplifted. The cap ranges from pale pink to rosy-red with lighter coloration at the edges. It is characteristically smooth, tacky/sticky, and is covered by a colorless gelatinous skin that can be peeled away. The **FLESH** within the cap is white, except just below the surface where it is pink. The flesh is thickest in the center and becomes abruptly thin at the edge. When cut the flesh darkens slightly if at all.

The **UNDERSIDE** of the cap has thick, waxy gills that are white when young, then turn smoky grey from spores, and finally become blackish when old. The gills are fairly well spaced and extend down onto the stem. The **SPORE** deposit is blackish. In young specimens the gills and stem are sheathed in a thin slime veil, which leaves a colorless ring near the top of the stem that later blackens from spores. The **STEM** is solid (not hollow). It is roughly equal in thickness throughout its length, or tapers (sometimes abruptly) towards the base. Above the ring the stem is silky white; below, it is moist, sticky or slimy. The lower quarter of the stem has flesh that is characteristically bright yellow throughout.

SIZE: Cap is 4–6 cm broad; stem is 3½–7(10) cm long and 5–15 mm thick.

LOOK-ALIKES: The common and edible *Gomphidius smithii* differs in that the cap is purplish grey and the lower part of the stem does not have yellow flesh, except sometimes at the extreme end. The edible but poor-tasting *Gomphidius oregonensis* differs in that the cap is dingy-salmon in color and it grows in clusters where the stems are fused at the base. See also the edible *Gomphidius glutinosus (p. 29)*.

OCCURRENCE: In the fall and winter this mushroom appears alone, in scattered groups or rarely in clusters (but not with fused stems). It grows in partnership with conifers, especially **spruce** and **Douglas-fir**. It is widely distributed and common.

COMMENTS: The flavor and odor are mild and not distinctive. This species is considered a mediocre to poor edible due to its bland taste and slimy texture. Remove the skin of the cap before cooking. The best results can be obtained by slicing the mushroom into thin strips and drying. The dried mushroom can then be used for cooking, or crushed into a powder for use as a seasoning agent. A degree of caution is warranted when eating larger amounts of this species, since it has a tendency to bioaccumulate heavy metals.

Slimy Spike Cap

syn. Hideous Gomphidius

- cap is cinnamon/violet and slimy
- gills extend onto stem
- flesh within stem base is yellow
- spore deposit is black

DESCRIPTION: The **CAP** is initially convex with edges that are inrolled. In maturity it expands to become broadly convex to flat, with the edges of the cap sometimes becoming uplifted. The cap is cinnamon brown to greyish-violet and is paler at the edges. When old it may develop dark streaks or stains. The upper surface is smooth and sticky/slimy. The **FLESH** within the cap is white and does not discolor when cut.

The **UNDERSIDE** of the cap has closely-spaced, waxy gills that extend fairly far onto the stem. Not present are short gills ("lamellulae") that only extend part of the way from the edge of the cap to the center. The gills are initially white and over time darken to smoky grey. The **SPORE** deposit is blackish. In young specimens the gills and stem are sheathed in a thin slime veil that leaves a colorless ring and a slight neck near the top of the stem. The ring later becomes black from spores. The **STEM** is solid (not hollow) and tapers slightly at the base, if at all. Above the ring the stem is white; below it is whitish to pale brown. The flesh in at least the lower half of the stem characteristically discolors to bright yellow when cut.

SIZE: Cap is 3–10(15) cm broad; stem is 3–10 cm long and 1–2 cm thick.

LOOK-ALIKES: The edible and common *Gomphidius smithii* differs in that it is smaller (3–6 cm broad) and the lower part of the stem does not have bright yellow flesh, except sometimes at the extreme end. The edible but less common *Gomphidius largus* is sometimes considered a variety of this species. It differs in being larger (up to 20 cm broad). The edible but poor-tasting *Gomphidius oregonensis* differs in that the cap is dingy salmon in color and it grows in clusters where the stems are fused at the base. See also the edible *Gomphidius subroseus (p. 28)*.

OCCURRENCE: In the summer and fall after rainy periods this mushroom emerges either alone, growing in scattered groups, or rarely growing in close clusters (but not with fused stems). It can be found in grass or needle litter under conifers, especially **spruce**, often in coastal forests. It is not common.

COMMENTS: The odor is mild and the flavor is not distinctive or is slightly acidic and/or sour. This species is considered a mediocre to poor edible due to its bland taste and slimy texture. Caution is also warranted since this species has a tendency to bioaccumulate heavy metals. Remove the slimy skin from the cap before cooking.

Horse Mushroom

- veil exhibits cogwheel pattern
- flesh does not discolor when cut
- cap smells of almonds or anise
- grows in grass

DESCRIPTION: The **CAP** is initially convex with inrolled edges, then becomes broadly convex to nearly flat. In maturity the edges of the cap may retain hanging remnants of the partial veil. The cap color is creamy white. The surface is smooth or sometimes slightly scaly in the center. The **FLESH** within the cap is white, firm, and does not discolor when cut. When young the mushroom smells of almonds or anise.

The **UNDERSIDE** of the cap has crowded gills that are free from the stem. The gills are initially white, then grey and finally turn dark chocolate brown from spores. The **SPORE** deposit is dark purple brown. In young specimens the gills are covered by a double veil that exhibits a characteristic cog-wheel pattern of scales around the stem. When the veil falls away it leaves a large, persistent, flaring double ring around the stem. The **STEM** is roughly equal in thickness throughout its length and may be pithy or hollow in the center. In maturity it is smooth above the ring and has small cottony patches below. The color of the stem ranges from creamy white to slightly yellowish. The flesh within the stem is creamy white and does not discolor when cut.

SIZE: Cap is 7–20 cm broad; stem is 5–15 cm long, 1–3 cm thick.

LOOK-ALIKES: A number of a potentially poisonous white mushrooms are superficially similar but lack the distinctive cog-wheel pattern on the veil. The edible *Agaricus silvicola* is very similar in appearance. It differs in that the stem is more slender and it grows in forest habitats rather than in grass. It has been known to cause allergic reactions in individual cases.

OCCURRENCE: In the summer and fall this common mushroom appears either alone, scattered, or in groups with each mushroom separated, often forming fairy rings. It grows in grassy areas such as lawns, fields and meadows.

COMMENTS: Only specimens young enough to have a veil should be collected for food since the distinctive cogwheel pattern makes them less likely to be confused with potentially poisonous look-alikes. This mushroom has a mild, nutty flavor and can be used in most mushroom recipes.

Meadow Mushroom

syn. Field Mushroom

- young gills are bright, bubble-gum pink
- skin does not turn yellow when rubbed
- stem base is not bulbous
- spore deposit is chocolate brown

DESCRIPTION: The **CAP** is initially convex with edges that are inrolled, then becomes broadly convex in maturity. The edges of the cap overhang slightly and often retain cottony remnants of the veil. When old, the cap flattens and sometimes becomes slightly depressed in the center. The cap is whitish to pale grey-brown and does not become yellowish when rubbed. When old it darkens at the edges. The surface is initially smooth, but may develop pale brownish scales in the center or become cracked like dried mud. In age the skin sometimes splits from the flesh at the edges of the cap, giving it a shingled appearance. The **FLESH** within the cap is thick and white. It does not discolor when cut.

The **UNDERSIDE** of the cap has crowded gills that are free from the stem. The gills are characteristically bright pink (like bubble gum) when young, then greyish pink when mature, and finally become chocolate brown or blackish from spores. The **SPORE** deposit is dark chocolate brown. When young, the gills are covered by a thin, white, partial veil that soon collapses and leaves a single ring on the stem. The ring is thin, white, and typically poorly-formed; in some cases most of the veil adheres to the edge of the cap rather than forming a ring. The **STEM** is white and usually tapers slightly towards the base. Its surface is smooth above the ring and thinly covered with woolly tufts below. The stem is initially stuffed with loose tissue but becomes hollow in age. As it ages the stem becomes pinkish and finally dull vinaceous brown in age. When bruised, it does not discolor to yellow. The flesh within the stem is firm and white when young, then becomes vinaceous brown in age. The base of the stem is not bulbous.

SIZE: Cap is 3–8(11) cm broad; stem is 2–6(10) cm long and 1–1½(2½) cm thick.

LOOK-ALIKES: A number of inedible or poisonous white mushrooms are superficially similar, but lack the bright pink gills when young. The edible but not-recommended *Agaricus semotus* has pinkish gills but differs in that it is smaller (2–7 cm broad) and the cap discolors to yellow when bruised. It causes gastrointestinal distress in some individuals.

OCCURRENCE: In the fall or in cool, rainy periods of the summer, this common mushroom emerges either alone, scattered, or in groups on the ground, sometimes forming arcs or rings. It grows in open, grassy areas, such as in meadows, fields and lawns, as well as along roads.

COMMENTS: This mushroom is considered a choice edible, and is closely related to the white grocery store mushroom. It has a mild, pleasant taste and slight odor that is reminiscent of almonds. Only eat meadow mushrooms that are young enough to still have the distinctively pink gills, in order to avoid confusion with inedible or poisonous look-alikes.

King Stropharia

syn. Wine Cap Stropharia, Garden Giant,
Burgundy Mushroom

- cap is wine-red
- ring has a cogwheel or claw-like pattern
- gills are whitish, then purplish gray
- spore deposit is purplish black

DESCRIPTION: This potentially large mushroom has a **CAP** that is initially convex, then becomes broadly convex to nearly flat in maturity. The edges of the cap sometimes retain ragged, hanging remnants of the partial veil. The cap color is variable. When young, it is characteristically wine-red or purplish brown and is irregularly covered with whitish veil remnants. Over time it may fade to tan or straw-color. The surface is initially smooth and sometimes slightly sticky, but is soon glossy and dry. When old, the surface sometimes becomes cracked. The **FLESH** within the cap is firm and white. It does not discolor when cut.

The **UNDERSIDE** of the cap has somewhat crowded gills that are either attached to the stem or have a notch where they attach. In age the gills may become free from the stem. Between the gills are frequent short-gills that extend only part of the way from the edge to the center. The gills are initially whitish, then gray, and finally purplish gray or purplish black with whitish edges. The **SPORE** deposit is dark purplish brown to purplish black. When young, the gills are covered by a white, membranous veil. The veil leaves a thick, well-developed ring on the stem. The ring is white to yellowish. The upper surface of the ring is finely grooved and often blackened by spores. The underside of the ring is radially split or "cogwheeled" into recurved segments that can appear claw-like. The **STEM** is stout, smooth and may be enlarged at the base. It is initially white, then discolors to yellowish and finally turns brownish when old.

SIZE: Cap is 4–15(20) cm broad; stem is 7–12(25) cm long and 1–3(7) cm thick.

LOOK-ALIKES: Specimens with purplish red caps have fewer look-alike species than tan-capped specimens. For that reason, only the former are recommended for eating. A number of species of *Agaricus* can have reddish caps, but differ in that they do not match all the key characteristics, including color of gills and color of spores.

OCCURRENCE: In the spring, summer and fall this mushroom appears either scattered or in larger numbers (sometimes in clusters) on the ground. It often appears on wood chips and leaf mulch in urban habitats, such as in lawns, gardens, and other cultivated areas. It may also occur along stream beds where spring floods have occurred.

COMMENTS: This mushroom has a pleasant but non-distinctive flavor and an odor that is faintly starchy.

Fairy Ring Mushroom

syn. Scotch Bonnet

- cap has central hump and pliable flesh
- gills are well-spaced
- stem is too tough to break with fingers
- grows in rings in grassy areas
- spore deposit is white

DESCRIPTION: The **CAP** is initially convex with inrolled edges. In maturity the cap broadens but retains a central hump that makes it flying-saucer shaped. The upper surface is smooth and ranges in color from creamy-brown to tan (but not whitish). In age the cap becomes paler and sometimes upturned at the edges. The cap is pliable and almost rubbery to the touch. The **FLESH** within the cap is and whitish to buff.

The **UNDERSIDE** of the cap has well-spaced gills that are interspersed with shorter gills ("lamellulae"). The gills are the same color as the cap and may be broadly attached, narrowly attached, or free from the stem. The **SPORE** deposit is white. The thin, smooth **STEM** is equally thick throughout its length or tapers slightly towards the base. The stem is solid (not hollow) and is characteristically pliant and tough; it cannot be broken with fingers or be easily separated from the cap. The upper portion of the stem is the same color as the cap and the lower potion is a darker shade of brown. When uprooted, the stem pulls up a fuzzy clump of mycelium and dirt.

SIZE: Cap is 2–5 cm broad; stem is 2–8 cm long and 3–5 mm thick.

LOOK-ALIKES: The poisonous *Clitocybe dealbata* also grows in grassy areas, also forms fairy rings, and also has a white spore deposit. It differs in that the cap is whitish and lacks a central hump, the gills are more closely spaced, and the stem is hollow and stiff rather than solid and pliant/tough. A number of other species are superficially similar but lack the distinctively tough and pliant stem and/or do not form fairy rings in grass.

OCCURRENCE; In the summer and fall these common mushrooms emerge in numbers, forming arcs or rings in the grass that are called "fairy rings." Because they are saprobic on grass, they occur in lawns, pastures and in parks, including areas where people walk frequently.

COMMENTS: This mushroom has a flavor that is unusually sweet and slightly nutty and an odor that is reminiscent of sawdust. Although considered a passable as an edible, these mushroom have a texture that may not appeal to everyone. They are a good candidate for drying and reconstitute well without lengthy soaking. When dried and powdered, they can be added to soups or even used in cookie recipes as a sweetener, due to high concentrations of the sugar trehalose. Use scissors to cut off and discard the extremely tough stems before cooking the caps.

Western Cauliflower Mushroom

- resembles cauliflower or sponge
- grows under conifers
- flesh is pliant, cartilaginous
- edges are flattened, wavy

DESCRIPTION: The fruiting body consists of flattened branches and wavy leaves that are packed together, forming a rounded, irregularly shaped mass. The **OUTSIDE** appearance is reminiscent of a head of cauliflower, a brain, or a sea sponge. In favorable conditions this mushroom can grow to gigantic proportions. The color is initially creamy white to greyish, then pale yellow to tan in maturity. It gradually darkens at the edges as it ages and does not discolor from handling.

The **BRANCHES** are short and thinly flattened with wavy, leaf-like, crisped ends that constitute the spore-bearing surface. The **SPORE** deposit is white. The branches arise from a large, root-like, sterile base that reveals itself to be chambered in cross-section. The base becomes progressively more solid as it tapers into the ground, where it has a deep underground portion. The **FLESH** within the base is pliant, cartilaginous, and white.

SIZE: Fruiting body is 10–60 cm broad.

LOOK-ALIKES: None.

OCCURRENCE: This mushroom usually appears in the late summer and fall after rainy periods. It grows alone and tends to recur annually in the same location. It feeds parasitically on dead and decaying roots at the base of trees and occurs primarily with conifers, especially **Douglas-fir** and various pines.

COMMENTS: Considered a good edible, the cauliflower mushroom has a fragrant, somewhat spicy odor and a sweet flavor. Leave the basal section in the ground when harvesting in order to encourage growth the following year. Be sure to thoroughly clean any debris from between the branches. This mushroom is one of the largest edible fungi, though smaller and younger specimens are preferred for cooking. For best results, slow-cook to soften the tissue. It can be used in soups or parboiled and then added as an ingredient to other dishes.

Orange Peel Fungus

syn. Orange Cup Fungus

- inside of cup is bright orange
- cup does not bruise or discolor
- no stem is present

DESCRIPTION: The fruiting body is cup- or saucer-shaped. As it expands, the cup becomes more flattened and eventually develops splits on the edges. It often grows pressed against adjacent neighbors, causing the cup to become infolded or develop wavy edges and an irregular shape. Occasionally the cup turns inside out to become convex, making it resemble an orange peel even more closely. The **INSIDE** (upper) surface of the cup is the spore-bearing surface. It is characteristically bright orange, smooth, and waxy/shiny.

The **OUTSIDE** (underside) of the cup ranges in color from pale orange to yellowish orange. It is smooth and when young is covered in a fine, white down. Neither the inside nor outer surface changes color when bruised. The thin layer of **FLESH** within the wall of the cup is light orange and brittle. The **SPORE** deposit is colorless. No **STEM** is present; the cup attaches to the soil by fungal threads.

SIZE: Fruiting body is 3–7(10) cm broad and 2–4 cm high.

LOOK-ALIKES: The poisonous *Otidea leporina* differs in that it is brownish rather than bright orange, is taller (to 7 cm high), has inward-folding edges, and has a cleft down one side. The inedible *Otidea onotica* differs in that it is orangey yellow rather than bright orange, is taller (up to 10 cm high), and has a cleft down one side. The inedible *Caloscypha fulgens* differs in that it is smaller (under 4 cm broad) and turns blue/green when bruised or old, particularly along the edges. The edible *Aleuria rhenana* differs in that it is much smaller (1–2 cm broad) and has a whitish rudimentary stem.

OCCURRENCE: In the spring and fall these cups appear on the ground after rainy weather, growing either alone, scattered, or in fused clusters. They occur on disturbed ground and are usually found on bare soil or in sandy/gravelly areas such as along trails and dirt roads. They also sometimes appear in grass or moss. This species is considered common.

COMMENTS: This colorful cup fungus is one of the few that is edible. When cooked it is not particularly flavorful. It is sometimes used as a garnish and can also be candied into a sugary treat. The process of candying involves boiling the mushroom and then steeping it in increasingly strong sugar solutions before drying off any remaining water. The high sugar saturation prevents spoilage.

35

Black Landscape Morel

- cap has ladderlike pits and ridges
- ridges are darker than pits in maturity
- cap and stem form a single hollow chamber
- grows in urban or landscaped areas

DESCRIPTION: This relatively large morel has a **CAP** that is typically elongated-conical though it is sometimes egg-shaped. The surface is convoluted with vertically oriented pits and ridges, as well as short horizontal cross-ridges, giving it a ladder-like appearance. When young, both the ridges and pits are pale to dark gray or brownish. In maturity the ridges darken more than the pits. The ridges become dark brown then nearly black, whereas the pits become dark gray, grayish brown, grayish olive, or brownish yellow. The exterior of the cap is the spore producing surface. The **SPORE** deposit is ocher/buff. Cutting the cap in half reveals a hollow interior, which is a key characteristic of morels. The **FLESH** of the cap is no more than a few millimeters thick. It is brittle or slightly rubbery and ranges in color from whitish to tan. The inner surface of the cap is whitish and pubescent.

The bottom of the cap does not hang free but rather is fused to a hollow stem, which together form a single, continuous, hollow interior. Where the edge of the cap meets the stem is a small groove-like sinus, which is 1–2 mm deep and wide. The **STEM** is cream to pale tan and is often somewhat swollen and pinched at the base. The surface of the stem is initially smooth, then often finely granulated in age. In maturity the stem develops longitudinal ridges and grooves near the base.

SIZE: Cap is 3–15 cm tall and 2–9 cm wide; stem is 3–10 cm long and 2–6 cm thick.

LOOK-ALIKES: The poisonous **false morels** (various species of *Gyromitra* and *Helvella*) as well as the not-recommended *Verpa bohemica* can have a similar appearance but differ in that the cap and stem do not form a single, continuous, hollow chamber. A number of other true morels (species of *Morchella*) are similar in appearance but differ in that they occur in natural rather than urban settings. All true morels are edible when cooked.

OCCURRENCE: In the late winter and early spring this fairly common morel appears either alone, scattered, or in numbers on the ground. It is limited to growing in urban environments, including in wood chips, gardens, and planters, and garden/landscape settings.

COMMENTS: This morel is considered a choice edible. It has an indistinct flavor and odor. Because raw morels are poisonous, be sure to cook them well. The scientific name of this species translates as "inconsiderate," which refers to its habit of appearing unexpectedly in urban areas.

Western Natural Black Morel

- cap is convoluted with pits and ridges
- ridges are darker than pits at all stages
- cap and stem form a single hollow chamber
- grows in non-burned hardwood forests

DESCRIPTION: The **CAP** is conical or nearly so and has a surface that is convoluted with ridges and pits. The ridges and pits are elongated vertically, though there are also frequent sunken, horizontal cross-ridges. When young, the ridges are dark brown to black and the pits between the ridges are pale tan. Over time the ridges become progressively darker, often turning black, and the pits also darken, becoming brown or yellowish brown. The cap exterior constitutes the **SPORE**-bearing surface. Cutting the cap in half reveals a hollow interior, which is a key characteristic of morels. The **FLESH** within the cap is 1–2 mm thick and whitish. The inner surface of the cap is whitish and pubescent.

The bottom of the cap does not hang free but rather is fused to a hollow **STEM**, which together form a single, continuous hollow interior - a key characteristic of morels. Where the edge of the cap meets the stem is a shallow trough, no more than a couple millimeters deep and wide. The stem surface is whitish, finely mealy, and lacks prominent ridges or folds. The base of the stem may be slightly swollen.

SIZE: Cap is 2½–3½ cm broad and 3–5 cm high; stem is 2–3½(5) cm long and 1–2(3) cm thick.

LOOK-ALIKES: The poisonous **false morels** (various species of *Gyromitra* and *Helvella*) as well as the not-recommended *Verpa bohemica* can have a similar appearance but differ in that the cap and stem do not form a single, continuous, hollow chamber. All true morels (species of *Morchella*) are edible when cooked. A number of such morels are similar in appearance but differ in that they occur in burned forest areas ("black burn morels"). The edible and somewhat similar-looking *Morchella populiphila* grows in forests that have not been burned. It differs in that the bottom of the cap hangs free from the stem. See also *Morchella importuna (p. 36)* and *Morchella snyderi (p. 38)*, both of which have dark ridges on their caps in maturity.

OCCURRENCE: In the spring this morel appears growing singly on the ground. Unlike "burn" morels, which occur only in areas that have been burned in intense wildfires, this morel fruits in non-burned forests. It grows in association with hardwoods and to a lesser degree conifers.

COMMENTS: As with other morels, this species is considered a choice edible, but should be thoroughly cooked in order to avoid adverse reactions. Light cooking or blanching is not sufficient. Morels preserve well by drying, which improves their flavor.

Thick Stemmed Morel

- cap is convoluted with pits and ridges
- ridges initially yellowish, then darken
- cap and stem form one hollow chamber
- stem is coarse and has folds
- grows in non-burned montane forests

DESCRIPTION: The **CAP** is conical when young, then becomes broadly conical in maturity. The cap surface is convoluted with pits and ridges. The pits and ridges are generally arranged vertically, with occasional sunken cross-ridges that are horizontal. The overall cap color is variable. The ridges and pits are initially yellowish. In maturity the ridges darken in color to become brown, gray, or black and the pits become tan to grayish brown. Because the color transition does not occur evenly, it's common to find specimens where the ridges are still yellowish towards the bottom on the cap, but have already begun to darken and turn black above. Cutting the cap in half reveals a hollow interior. The **FLESH** of the cap is whitish.

The bottom of the cap does not hang free but rather is fused to the hollow stem, forming a single, continuous, hollow interior. Where the edge of the cap fuses to the stem is a sinus, which is a cavity a few millimeters deep and wide. The **STEM** is characteristically folded and pocketed, or "lacunose," which is a key characteristic of this species. It is whitish, tan or rose-tinted, and becomes pale brownish in age. Its surface texture is initially mealy with whitish granules, then becomes prominently granulated and coarse. The base of the stem may be slightly swollen.

SIZE: Cap is 3½–8 cm tall and 3–5 cm wide; stem is 3½–7 cm tall and 2½–4 cm thick.

LOOK-ALIKES: The poisonous **false morels** (various species of *Gyromitra* and *Helvella*) as well as the not-recommended *Verpa bohemica* differ in that the cap and stem do not form a single, continuous, hollow chamber. All true morels (species of *Morchella*) are edible when properly cooked. A number of other true morels can have a similar appearance, and are sometimes difficult to differentiate in the field, but do not generally match all the key characteristics. See also *M. brunnea (p. 37)* and *M. tridentina (p. 40)*, which also grow in non-burned forests.

OCCURRENCE: In the spring this common morel appears either alone, scattered, or growing in tight clusters where the stems are fused. It occurs in montane regions, where it can be found on sand, soil, duff or in forest litter under conifers.

COMMENTS: As with most other morels, this species in considered a choice edible but must be thoroughly cooked to avoid illness; blanching or light simmering is not adequate. Morels have a mild flavor and earthy odor. They preserve well by drying, which concentrates the flavor.

Blushing Morel

- cap is convoluted with pits and ridges
- ridges are lighter than pits at all stages
- cap bruises reddish/orange
- no trough exists where cap fuses to stem
- grows in urban areas

DESCRIPTION: The **CAP** is conical to narrowly conical and has a surface that is convoluted with pits and ridges. In young specimens the pits and ridges are somewhat vertically arranged and have irregular horizontal cross-ridges. As the cap develops, the ridges and pits become more irregular, and exhibit less vertical alignment. Young specimens have brown pits with ridges that are paler (beige to yellowish brown). In maturity the pits lighten in color to become yellowish brown. When bruised, the cap slowly discolors to orange or reddish brown. The **FLESH** within the cap is whitish to tan, and is firm to slightly rubbery.

The bottom of the cap does not hang loose but rather is fused to the stem, forming a single, continuous, hollow interior. Unlike many other morels, there is no sinus or trough where the edge of the cap fuses with the stem. The cap exterior constitutes the spore-bearing surface. The **SPORE** deposit is pale orange. The **STEM** is rigid, hollow, and has a smooth surface. It ranges in color from off-white to light brown. The base of the stem may widen and be pinched.

SIZE: Cap is 2–7(12) cm tall and 2–5 cm broad; stem is 3–6(9) cm tall and 1–3 cm thick.

LOOK-ALIKES: The not-recommended *Verpa bohemica* differs in that it does not discolor to orange when bruised and the bot-

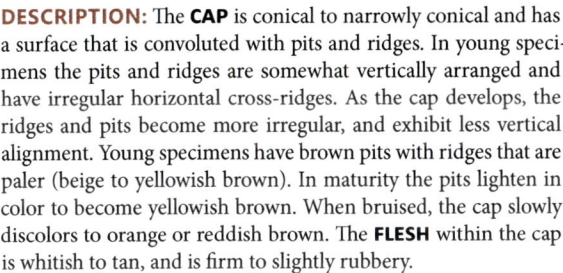

tom of the cap hangs loose from the stem rather than forming a single, continuous, hollow chamber. All true morels (species of *Morchella*) are edible when properly cooked. *Morchella americana* has a similar, light coloration, but differs in that it does not turn reddish when bruised and the pits and ridges are not vertically aligned at any stage of development. See also *Morchella tridentina (p. 40)*, which differs most obviously by growing in montane forests.

OCCURRENCE: This species appears in the spring and summer, either alone or more typically in groups, often as the last of the morels to emerge. It occurs near riverbanks as well as in old orchards, disturbed soil, irrigated gardens and in woodchips.

COMMENTS: This species is considered a good edible but is not as highly prized as other morels. Like other morels, it contains a small amount of toxin that is destroyed by thorough cooking. When collecting this species for food, bypass those that are growing in wood chips where the wood itself may have been treated with toxic chemical agents.

Mountain Blond Morel

syn. Western Blond Morel

- cap is convoluted with pits and ridges
- ridges are lighter than pits at all stages
- ridges and pits are oriented vertically
- groove exists where cap fuses to stem
- grows in non-burned montane forests

DESCRIPTION: The **CAP** is conical and has a surface that is convoluted with pits and ridges. The ridges and pits are vertically aligned and intersect with sunken, horizontal cross-ridges. The ridges are smooth, yellowish tan to yellowish brown and remain pale through to maturity. Between the ridges are smooth pits, which are typically elongated vertically. When young, the pits are whitish to dull gray or pale yellow, then pale tan or pale pinkish tan in maturity. When bruised, the cap may "blush," discoloring to orange. Cutting the cap in half reveals a hollow interior with an inner surface that is whitish and pubescent. The **FLESH** of the cap is 1–3 mm thick and whitish.

The bottom of the cap does not hang free but rather is fused to the stem, forming a single, continuous, hollow chamber. Where the cap attaches to the stem is a groove that is 2–4 mm deep and wide. The cap exterior constitutes the **SPORE**-bearing surface. The **STEM** is more or less equally wide along its length, though it may be thicker at the base. The surface is white and has a texture that is either smooth or finely mealy with whitish granules, except at the base, where it is smooth.

SIZE: Cap is 4–6 cm high and 2½–4 cm broad; stem is 2–6 cm long and 1–4 cm thick.

LOOK-ALIKES: The not-recommended *Verpa bohemica* differs in that the bottom of the cap hangs loose from the stem rather than forming a single, continuous, hollow chamber. All true morels (species of *Morchella*) are edible when properly cooked. *Morchella americana* has a similar, light coloration but differs in that the pits and ridges are not vertically aligned and instead have a random orientation. In addition, it lacks a groove-like sinus between the stem and cap. See also *Morchella rufobrunnea (p. 39)* and *Morchella snyderi (p. 38)*, which is initially yellowish.

OCCURRENCE: In the spring, this morel appears either alone, scattered, or in small groups on the ground. It occurs in montane regions, growing in non-burned conifer forests.

COMMENTS: This morel is considered a choice edible, though it must be thoroughly cooked and never eaten raw. It has a flavor that is milder than that of black morels.

Bell Morel

syn. Thimble Fungus, Smooth Thimble Cap

- cap is thimble-shaped
- cap hangs free from stem
- stem is loosely stuffed with cottony fibers
- spore deposit is yellow

DESCRIPTION: Despite its common name, this mushroom is not a "true" morel (i.e., species of *Morchella*), though like a morel it is well worth collecting. The **CAP** is bell-shaped with edges that are initially incurved. Unlike a true morel, the cap is attached only at the top of the stem, like a thimble hanging from the top of a pencil. The bottom edges of the cap hang free. In maturity the edges may flare outwards or even become slightly upturned. The cap color ranges from honey brown or ocher brown to dark brown. The surface is typically smooth or slightly wrinkled, but may be broadly convoluted. It is tacky when moist and may exhibits faint furrow lines near the edges. The **SPORE** deposit is yellow. The **FLESH** within the cap is white, thin and brittle.

The **INNER SURFACE** of the cap ranges in color from tan to dark brown. The **STEM** is equally wide throughout its length and is usually round in cross section. It has a surface that is white and smooth, but may exhibit small orangey to brownish granules that form transverse belts or ribs. The inside of the stem is loosely stuffed with cottony fibers and becomes hollow when old.

SIZE: Fruiting body is 3–13 cm high; cap is (1)1½–2(4) cm broad; stem is (2½)4–11(13) cm long and 5–15 mm thick.

LOOK-ALIKES: The poisonous *Gyromitra esculenta (p. xxi-i)*, as well as other poisonous or not-recommended species of *Gyromitra*, differ in that the cap does not hang on the stem like a thimble from the top of a pencil. The not-recommended *Verpa bohemica* differs in that the cap is convoluted with pits and ridges, giving it a more strongly wrinkled appearance. True morels *(pp. 40-41)*, which are edible when thoroughly cooked, differ in that the cap is strongly convoluted with pits and ridges and the cap does not hang on the stem.

OCCURRENCE: In the spring this mushroom appears either alone, widely scattered, or growing abundantly in groups on the ground. It occurs near streams and riverbanks as well as in orchards, along hedges, and in hardwood forests or in mixed woods, where it grows in the humus of the forest floor.

COMMENTS: This mushroom has a pleasant but non-distinctive flavor and odor. Due to its insubstantial size and hollow nature, a good number must be collected in order to have enough for a meal. These mushrooms are sometimes parboiled and the water discarded before using the mushrooms in a final preparation.

41

Gem-studded Puffball

syn. Common Puffball, Devil's Snuff-box

- has conical spines when young
- interior is all-white, undifferentiated
- spines rub off, leaving white scars

DESCRIPTION: This puffball has a variable form; it can resemble a spinning top, pestle, or an inverted egg or pear. There is often a slight bump at the top, where a small hole eventually develops for dispersing spores. The **SURFACE** is white to pale brown. When young it is covered with tiny conical spines, especially on the upper portions. In maturity the spines fall off, leaving a mesh-like pattern of round scars.

The **INTERIOR** or "gleba" is firm and all-white when young. In maturity, the gleba becomes yellowish to olive brown and powdery. The **SPORE** color is brown, though by the time spores are present the puffball is long past being edible. The **STEM**-like base is continuous with the rest of the fruiting body. It is about half as wide, and constitutes up to two thirds of the height of the puffball. It is similar in color but has few or none of the spines/scars. The base is often folded or marked with parallel grooves that run to the bottom. The flesh within the base is initially white, then becomes grey-brown in maturity.

SIZE: Fruiting body is 2–6(9) cm broad and 2–7(10) cm high; spines are 1–2 mm long.

LOOK-ALIKES: A number of potentially poisonous, white mushrooms are similar in appearance when they are in their immature button stage. Button-stage mushrooms differ in that they are smaller (under 4 cm broad) and exhibit the faint outline of gills or other structures when cut in half, rather than having undifferentiated flesh. The poisonous **earth balls** (species of *Scleroderma*) are superficially similar, but are tan or yellowish to brown rather than white, lack spines, and have a black interior (though when very young, the interior is whitish). The not-recommended *Lycoperdon marginatum* differs in that the spines group together and join at the tips and do not shed individually but rather peel away in patches or sheets. The probably edible *Vascellum curtisii* differs in that it is smaller (under 2 cm broad) and has longer spines (up to 5 mm long) that join together at the tips to form spikes, though the spines often fall away. Species of white puffball that lack spines at any stage, including *Calvatia cyathiformis* and *Bovista plumbea*, are generally edible if the interior is all-white.

OCCURRENCE: In the summer and fall, this common puffball appears on the ground sometimes alone but more often scattered or in clusters. It can be found in conifer or mixed forests and less commonly in cultivated or grassy areas and along roads.

COMMENTS: As with other edible puffballs, this species cannot be eaten after the interior turns from white to yellowish. It has a flavor that is mild to slightly bitter, and a texture that is foamy and marshmallow-like.

Stump Puffball

syn. Pear-shaped Puffball, Wolf-fart Puffball

- grows in clusters on wood
- interior is all-white and undifferentiated
- attached to wood by fine, white rootlets

DESCRIPTION: When young, this puffball is more or less round. In maturity it remains round or becomes shaped like an inverted pear. The **SURFACE** color is initially pale tan to light brown. Over time it becomes either yellowish brown or darkens to rusty brown. Young puffballs have a surface texture that is generally smooth but may have a scattering of tiny whitish spines and granules. After the spines disappear, the outer layer eventually becomes finely cracked like dried mud. The cracks form smaller and smaller patches that dry out and become a rough, granular surface. Beneath the outer layer is a papery, firm inner layer that is initially white, then becomes light brown to chestnut brown.

The **INTERIOR** or "gleba" is white and fleshy when young. At this stage the puffball is edible. Over time the interior turns yellowish, then olive, and finally dries out to become olive brown **SPORE** dust. The top of the puffball eventually perforates to form an irregular vent. It is through this hole that spores are dispersed with the aid of rain drops and/or air currents. The bottom of the puffball often has a crumpled or folded/corrugated appearance where it pinches off to form the base. Whether the **BASE** is poorly developed or prominent affects the overall shape of the fruiting body. The flesh within the base is white and spongy when fresh, and it characteristically remains white into maturity. The puffball is attached to the wood substrate by numerous white, radiating, thread-like rootlets.

SIZE: Fruiting body is up to 3(5) cm high and wide; the base is ⅓ to ½ of the total height.

LOOK-ALIKES: A number of poisonous mushrooms may resemble this species when they are in their button stage. Such mushrooms differ from puffballs in that they exhibit the outline of gills and other structures when cut in half, rather than having an interior that is all-white and undifferentiated. In addition, a number of not-recommended or poisonous, terrestrial, brown puffballs are superficially similar in appearance. They differ most obviously in that they grow on the ground rather than from wood.

OCCURRENCE: In the fall this common puffball appears either scattered or (more often) in dense clusters on dead and decaying wood. It can be found growing on hardwood and conifer roots, stumps and logs, sometimes in the hundreds.

COMMENTS: This puffball is only edible when it is young and when the interior is all-white. At this stage it has a mild flavor and an odor that is strong and somewhat unpleasant. A popular way of cooking puffballs is by slicing them and then frying the slices in oil, with or without first battering them.

Sculptured Giant Puffball

syn. Warted Giant Puffball

- grows to grapefruit-size
- has thick skin with pyramidal warts
- grows in mountains under conifers

DESCRIPTION: This puffball is roughly spherical or irregularly top-shaped, like an inverted pear, and grows to be about as large as a grapefruit. The **SURFACE** color is whitish to cream or pale ochre, then light yellow-brown in age. It has a two-layered skin, or "peridium." The outer layer is thick, leathery, and has a texture that is felt-like. It characteristically develops broad, low pyramidal warts. The warts are irregularly three- to six-sided and are usually blunt, but sometimes pointed. The pyramids are grooved with parallel transverse lines. The warts diminish near the base, where the outer layer becomes thinner and smooth.

The inner layer of the peridium is thin and fragile. It is a shiny, membranous tissue that is depressed around the perimeter of each pyramidal plate. In maturity the peridium cracks open at the base of the warts, or the apex of the fruiting body peels back, exposing a powdery, spore-filled interior. In young specimens, the **INTERIOR** or "gleba" is initially white and firm. In maturity the gleba turns sulphur yellow as the spores mature, then golden brown/olivaceous, and finally dark umber brown to purplish brown as it becomes powdery. The **SPORE** color is brown, though by this time the puffball is long past being edible. Beneath the gleba is the "subgleba," a region that occupies the lower quarter to a third of the puffball. The subgleba consists of chambers of sterile tissue that persists after spore dispersal. The subgleba is initially cream colored, then buff to light brown or purplish. The **BASE** of the puffball has a surface that is often pleated into folds. The base is attached to the soil by a root-like mycelial cord.

SIZE: Fruiting body is 7–16 cm broad and up to 10 cm high; pyramids are raised 5–8 mm.

LOOK-ALIKES: A number of potentially poisonous, white mushrooms are similar in appearance when in their immature button stage. Button stage mushrooms differ in that they are smaller (under 4 cm broad) and exhibit the faint outline of gills or other structures when cut in half. The uncommon but edible *Calvatia booniana* differs in that it is larger (10–60 cm broad) and lacks a sterile base. See also *Calvatia sculpta (p. 45)*.

OCCURRENCE: In the spring and summer this common mountain puffball appears alone or less commonly scattered or in small clusters on the ground. It is a saprobic species, decomposing dead plant material. It occurs by roadsides, in open coniferous forest, or near the forest edge at subalpine or occasionally alpine elevations on the eastern side of the Cascade range, particularly under ponderosa pine.

COMMENTS: This puffball is edible when it is young, while the interior is still all-white. The taste and odor are not distinctive and it is not considered choice. As with other puffballs, it can be prepared for eating by slicing and frying it in butter or oil, with or without first battering.

Sculptured Puffball

syn. Sculpted Puffball, Sierran Puffball

- grows to fist-size
- has thick skin with pointed warts
- grows in mountains under conifers

DESCRIPTION: This puffball is roughly spherical or irregularly top-shaped, like an inverted egg or pear. The **SURFACE** is white and is composed of a two-layered skin, or "peridium." The outer layer, or "exoperidium," is thick and covered in large, distinctive warts that are pyramid-shaped or polygonal. The pyramids are long and pointed. Each pyramid may be erect or bent over, and may be joined at the tip with a neighboring pyramid. The tips of the pyramids coil and darken as they dry. The pyramidal surface is streaked with parallel horizontal lines that run either horizontally or longitudinally.

The inner layer of the peridium is a thin, fragile tissue that eventually breaks up with the expoperidium. In maturity the peridium opens by cracks formed at the base of the warts, though the warts may remain attached at the tips. Beginning with the upper surface, the pyramidal segments slough off when they are fully mature, creating an irregular opening that exposes the powdery, spore-filled interior cavity. In young specimens, the **INTERIOR** or "gleba" is initially white. As it matures the gleba changes color as the spores mature, turning yellowish, and finally deep olive-brown when powdery. The **SPORE** color is brown, though by the time spores are present the puffball is long past being edible. The region beneath the gleba is called the "subgleba." The subgleba consists of chambers of sterile tissue that persists after spore dispersal. The subgleba is often purplish in the base. The **BASE** of the puffball does not root.

SIZE: Fruiting body is (4)8–10 cm broad and 8–15 cm tall; pyramids are raised up to 3 cm.

LOOK-ALIKES: The probably edible *Vascellum curtisii* differs in that it is smaller (under 2 cm broad) and has much shorter spines (up to 5 mm long). See also the edible *Calbovista subsculpta (p. 44)*.

OCCURRENCE: This uncommon puffball appears throughout the spring, summer and fall in wet periods, growing either alone, scattered, or in small groups on the ground. It grows in conifer duff at high elevations (greater than 750 m or 2500 ft).

COMMENTS: This puffball is edible when it is young enough that the interior is firm and all-white. At this stage it has a mild taste, little to no odor, and is considered choice by some accounts. Older specimens may have an unpleasant iodine-like flavor. As with other puffballs, it can be prepared for eating by slicing and then sautéing it in butter or oil, with or without first battering it. It can also be preserved by freezing the fresh or partially cooked slices. This species was used as a traditional food of the Plains and Sierra Miwok Indians.

Hedgehog Mushroom

syn. Wood Hedgehog

- cap is tan, bruises orange
- underside is covered in spines
- stem is squat and paler than cap
- flesh is whitish

DESCRIPTION: The bristly underside of this mushroom earns it the moniker "hedgehog." The **CAP** is initially convex with inrolled edges and becomes irregularly flattened over time. In maturity it often develops a central depression, becomes wavy, and sometimes lobed. The cap color is tan with orange tones and it discolors to a deeper shade of orange when handled or bruised. The surface texture is smooth to slightly felt-like. The **FLESH** within the cap is firm, brittle and whitish. It discolors unevenly to dull orange when cut and sometimes exhibits color zones.

The **UNDERSIDE** of the cap is covered in spines or "teeth," which are the spore-bearing surface. The spines are brittle and hang down like stalactites. They are whitish to pale orange and discolor to a deeper orange when bruised. The spines may or may not extend partly down onto the stem. The **SPORE** deposit is white. The **STEM** attaches to the cap either centrally or off-center. It is solid (not hollow) and may narrow slightly in either direction. The stem is paler than the cap and bruises to a deeper shade of orange.

SIZE: Cap is 2–17(25) cm broad; stem is 3–10 cm long and 1–3(5) cm thick; spines are 2–6 mm long.

LOOK-ALIKES: The inedible *Hydnellum suaveolens* differs in that the stem is purplish or bluish and it has a strong odor of anise or peppermint. The inedible *Hydnellum caeruleum* differs in that the cap is bluish (more so when young), the stem is orange, and the spore deposit is brown. The inedible *Hydnellum aurantiacum* differs in that the cap is orange to orange-brown and the spore deposit is brown. *Bankera violascens*, of unknown edibility, differs in that the cap is brown with purple or grey tones and the stem is darker than the cap. See also *Hydnum oregonense (p. 47)*.

OCCURRENCE: In the summer and fall this mushroom emerges either alone, scattered, or growing in groups. It appears in the leaf litter and moss of forest floors, usually in association with conifers such as **Douglas-fir**, but also with hardwoods.

COMMENTS: This popular edible has a pleasant odor and a delicious flavor, though older specimens may be somewhat bitter or peppery if not well cooked. Because it is resistant to infestation, the older and larger specimens can often be harvested. The spines are typically removed before cooking. This mushroom is difficult to reconstitute after drying.

Sweet Tooth Mushroom

syn. Depressed Hedgehog

- cap is pale orange, bruises orange
- cap becomes umbilicate
- underside is covered in spines
- stem is narrow, centered, paler than cap
- flesh is whitish

DESCRIPTION: This species is the little brother of the **hedgehog mushroom** *(p. 46)*. It has a **CAP** that is initially convex with slightly inrolled edges. In maturity it becomes broadly convex or irregularly flat, often with wavy edges. The cap develops a central depression or indentation that becomes umbilicate (i.e., perforated into the stem). The cap color is tan to pale orange and slowly discolors to a deeper orange when bruised. The surface is smooth or slightly felt-like. The **FLESH** within the cap is whitish/tan and discolors to orange when cut.

The **UNDERSIDE** of the cap is covered with spines or "teeth" that hang down like stalactites. The undersurface is slightly depressed around the stem and the spines do not extend onto the stem itself. The spines are pale yellow and they discolor to pale orange when bruised. The **SPORE** deposit is white. The **STEM** is roughly equal in thickness throughout its length, though it may be slightly swollen at the base. The stem usually attaches to the cap centrally, but occasionally is off-center. It is paler than the cap and discolors to a deeper orange when bruised.

SIZE: Cap is 3–5 cm broad; stem is 3–8 cm long and under 1 cm thick; spines are 2–6 mm long.

LOOK-ALIKES: See also *Hydnum repandum (p. 46)*.

OCCURRENCE: In the summer and late fall this mushroom appears alone or in numbers. It grows in the duff of coniferous forests and conifer bogs, as well as in mixed forests.

COMMENTS: This mushroom has an odor that is mild and slightly fruity and a flavor that ranges from mildly sweet to slightly bitter. The spines are normally removed before cooking in order to keep them from scattering in the dish and/or overcooking. These mushrooms fare best when slow-cooked in order to make them more tender. They do not reconstitute well after drying, but can be preserved by freezing after partial sautéing.

Scaly Hedgehog

syn. Hawk's Wing, Shingled Hedgehog, Scaly Tooth, Scaly Hydnum

- cap has overlapping scales
- scales are coarse
- underside has brown teeth
- stem is brown

DESCRIPTION: The fruiting body has a **CAP** that is initially convex with inrolled edges. As it ages it flattens and then becomes depressed in the center. In maturity the edges becomes wavy and/or lobed. When old the center become perforated, funneling into a hollow part of the stem. The cap is initially velvety and light brown, then develops large, coarse scales that are dark brown to almost black. The scales overlap each other like shingles. When old they often turn up at the ends and may partially weather away. The **FLESH** within the cap is thick, brittle and creamy white to pale brown.

The **UNDERSIDE** of the cap is covered with spines or "teeth" that hang down like stalactites and constitute the spore-bearing surface. The spines are initially whitish, then become brown in maturity. They may extend slightly down onto the stem. The **SPORE** deposit is brown. The **STEM** is attached either centrally or off-center. It is roughly equal in thickness throughout its length, though it may be enlarged at the base. The stem is smooth and initially solid, becoming hollow in the top part when old. Its is initially whitish and becomes brown in maturity except at the base, where it is covered with white mycelium.

SIZE: Cap is 5–25 cm broad; stem is 4–8 cm long and 2–3½ cm thick; spines are ½–1 cm long.

LOOK-ALIKES: The similar-looking but inedibly bitter *Sarcodon scabrosus* differs in that it is smaller (under 10 cm broad), the cap has reddish or purple tones, and the stem base is olive-black. The inedible (or less favored) *Sarcodon squamosus* differs in that the cap is darker and has a center that is rarely depressed, the cap edges remain incurved, and it grows under pine.

OCCURRENCE: In the summer and fall this mushroom appears either alone, scattered, or forming arcs on the ground. It occurs in moss, duff, and on sandy soil, and grows in association with **Engelmann spruce** and **subalpine fir**. It is especially common in the Rocky Mountains.

COMMENTS: This mushroom has a slightly spicy odor and a mild flavor that is sometimes followed with a bitter aftertaste. It causes indigestion in some individuals. Young specimens are best for collecting, since they tend to be infested when older. When dried this mushroom has a smoky/chocolate odor.

Deer Mushroom

syn. Deer Shield, Fawn Mushroom

- grows on dead wood
- gills are whitish and free from the stem
- stalk is white with longitudinal striations
- smells of radishes
- does not bruise blue
- spore deposit is cinnamon

DESCRIPTION: The *Pluteus cervinus* group consists of a number of closely related and similar-looking, edible species. The **CAP** is initially bluntly convex. As it matures it becomes broadly convex or bell-shaped, then flattens out but typically retains a low, broad, central hump. The cap color is variable, ranging from pale brown to dark brown, with the center often darker than the edges. In maturity it may develop faint, radial streaks. The skin is smooth, slightly sticky when wet, and can be peeled from the edge to the central hump. The **FLESH** within the cap is soft and whitish, does not discolor when cut, and smells faintly of radishes.

The **UNDERSIDE** of the cap has broad, closely spaced gills that are characteristically free from the stem. The gills are initially whitish and become progressively flesh-colored as the spores mature. The **SPORE** deposit is cinnamon. The **STEM** is firm, solid (not hollow) and may widen slightly towards the base. It has a dingy appearance due to fine, greyish brown longitudinal fibrils and/or striations. No ring or veil is present.

SIZE: Cap is 3–12(15) cm broad; stem is 5–13 cm long and 5–20 mm thick.

LOOK-ALIKES: A number of poisonous or inedible brown mushrooms that grow on wood are similar in appearance but do not match all the key characteristics. For example, the deadly *Galerina marginata* (Fig. 9e) is similar in color but differs in that the cap is smaller (up to 4 cm broad), the stem has a ring, and the gills are not free from the stem. A large number of closely related species of *Pluteus* are similar in appearance. Of these several species bruise blue and are pychoactive due to the presence of psilocybin. The edible *Pluteus petasatus* differs in that the cap is cream colored with a brownish center. Because it often grows from buried wood rather than from stumps, it may be confused with poisonous terrestrial species (such as various species of *Entoloma*) and therefore is not recommended.

OCCURRENCE: In cool, damp periods in the spring and late fall, this common mushroom appears either alone or scattered in small groups, but not in clusters. It grows from decaying hardwood logs, stumps, or other woody debris.

COMMENTS: Members of this species complex range from mediocre/good to poor edibles. The flavor and odor are reminiscent of radishes. Fresh, firm specimens are best for eating, since older and softer mushrooms have a pappy texture.

Chicken of the Woods

syn. Chicken Mushroom, Sulphur Shelf, Sulphur Polypore

- cap has orange and yellow bands
- grows on **conifers**
- edges are yellow (edible if soft)
- underside is yellowish, porous

DESCRIPTION: This easy-to-spot fungus has long been a favorite of mushroom hunters. The fruiting bodies usually grow as overlapping shelves on conifers. The **CAP** is fan-shaped or semicircular with softly rounded edges that are often undulating. The upper surface exhibits bright yellow and orange banding, with the edges yellow. In age the colors become less intense. The texture is initially felt-like, then becomes finely wrinkled and radially furrowed in maturity. The **FLESH** within the cap is white to pale yellow. It is initially soft and spongy, then becomes tougher, brittle and chalk-like.

The **UNDERSIDE** of the cap, when examined closely, reveals tiny, round pores that are the openings of the tube layer. The pore surface is yellow and when young may exude a yellow juice if squeezed. The **SPORE** deposit is white. There is no **STEM**; the cap is attached laterally to the host tree.

SIZE: Cap is 5–30(50) cm broad and up to 3 cm thick; pores are 2–4 per mm.

LOOK-ALIKES: The not-recommended (but widely eaten) *Laetiporus gilbertsonii* grows on hardwoods rather than conifers, but otherwise has an identical appearance. It has been implicated in adverse reactions including nausea in some individuals. The inedible and uncommon *Pycnoporus cinnabarinus* differs in that it is red or reddish, is not banded with yellow, and the edge of the cap is not soft and undulating.

OCCURRENCE: From the late spring to early fall this mushroom appears either alone or more commonly in overlapping shelves. It grows on dead or living conifer trees, stumps, or logs and reappears year after year in the same location.

COMMENTS: This popular mushroom must be thoroughly cooked before being eaten. Even when well-cooked it causes gastrointestinal upset in some individuals. It is a meaty and delicious mushroom that sometimes has a mild lemony flavor. Soft, young specimens are the best for cooking. When older they become progressively tougher and/or more sour tasting, although the soft outer edge can sometimes be harvested with good results.

Oyster Mushroom

syn. Pearl Oyster

- cap is over 9 cm broad
- grows on hardwood trees
- gills descend all the way to base
- gills are not saw-toothed or ruffled
- spore deposit is grey/lilac

DESCRIPTION: Cultivated varieties of this mushroom often appear for sale in the local grocery store. The **CAP** is initially semicircular and convex with inrolled edges. In maturity it becomes fan- or kidney-shaped and nearly flat. Sometimes the center becomes slightly depressed and the edges become wavy, lobed or split. The cap color is a variable shade of light brown. When young the surface texture is moist and almost greasy but not slimy; later it is dry and smooth. The **FLESH** within the cap is thick, firm and whitish. It does not discolor when cut. The **UNDERSIDE** of the cap has broad gills that are fairly closely spaced and are **not** serrated/saw-toothed. The gills descend right to the base, without becoming ruffled/wavy at the base. There are also shorter gills ("lamellulae") that start at the edge and run only part of the way towards the base. The gills are **not** separable from the cap as a layer. The **SPORE** deposit is lilac/grey. There is **no** stalk-like **STEM**; the cap attaches laterally to the host tree with a short, white, lateral pseudostem.

SIZE: Cap is (4)9–15(25) cm broad; stem is ½–3 cm long and ½–2 cm thick.

LOOK-ALIKES: The inedible/poisonous *Tapinella atrotomentosa* differs in that it has a thick, sturdy stem (sometimes laterally attached), the stem base is covered in velvety brown fuzz, the gills form a separable layer, and it grows on conifer wood. *Tapinella panuoides*, of unknown edibility, differs in that the cap is smaller (2–7 cm broad), the gills often become ruffled or wavy near the base, the spore deposit is brownish, and it grows on conifer wood. A number of inedible species of *Lentinellus* are superficially similar in appearance but differ in that the gills are serrated. Other look-alikes that are inedible or of unknown edibility include various members of *Crepidotus, Panellus,* and *Pleurotopsis*; they differ in that they are relatively small (under 5 cm broad). The common name "oyster mushroom" also refers to several other edible species of *Pleurotus*. The edible *Pleurotus pulmonarius* and *Pleurotus dryinus* differ in that they have a stalk-like stem that is more centrally attached. The edible *Pleurotus populinus* differs in that it is smaller (under 9 cm broad), has a white spore print, fruits earlier in the spring, and grows only on **aspens** and **cottonwoods**. See also the edible *Panellus serotinus (p. 52)*.

OCCURRENCE: In the fall, winter and spring, this mushroom appears either alone or more often in clusters of overlapping shelves, sometimes in enormous numbers. It occurs on stumps and on fallen or standing trunks of dead/dying trees (usually hardwoods) – not on the ground.

COMMENTS: This well-known mushroom exudes an almond- or anise-like aroma. It has a mild, pleasant flavor and a slightly chewy texture. The cultivated supermarket variety has different coloration but similar flavor. Oyster mushrooms do not reconstitute well after drying.

Late Oyster

syn. Late Fall Oyster, Mukitake

- cap is green, purple, or yellow
- cap is sticky or tacky when wet
- grows on dead wood
- occurs late in the fall
- spore deposit is yellow

DESCRIPTION: The **CAP** is fan- or kidney- shaped and initially convex with incurved edges. In maturity it becomes nearly flat with edges that remain incurved. Over time the edges become wavy and usually develop lobes or indentations. The cap color is variable, and can be dull green, yellow, purple or a mixture of several shades. The thick skin is smooth and can become finely velvety in age. In moist conditions it is tacky or sticky. The **FLESH** within the cap is white, firm and rather tough but pliant.

The **UNDERSIDE** of the cap has narrow, closely-spaced gills that do not fork. They are broadly attached to the pseudostem and do not descend onto it. The gills are yellowish, orangey or greenish and become brown at the edges when old. The **SPORE** deposit is yellow. The **PSEUD-OSTEM** is laterally attached to the cap. It is yellow to orangey.

SIZE: Cap is 3–10 cm broad; stem is 1–2 cm long and 1-2 cm thick.

LOOK-ALIKES: Specimens that are greenish and tacky are best for collecting for food since they are unlikely to be confused with other species. Specimens that are not greenish (e.g., yellowish, tan, brownish, or purple) have a number of inedible or poisonous look-alikes, however these can be differentiated by comparing all the key characteristics. See also *Pleurotus ostreatus (p. 51)* and *Panus conchatus (p. 53)*.

OCCURRENCE: In the late fall this mushroom appears either alone or more often in clusters of overlapping shelves. It grows on the decaying wood of both conifers and deciduous trees, such as from the sides of logs and on fallen branches that still retain their bark. It does not decay rapidly and persists for some time. It is considered uncommon.

COMMENTS: This mushroom has a mild odor and a variable flavor that ranges from delicious to bitter. The flavor may be influenced by the host tree. Although it causes indigestion in some individuals, it is popular in Japan, where it is called **mukitake**. For best results when cooking, slice the mushroom thinly and thoroughly dry sautée in order to eliminate as much moisture as possible. Alternatively, bake in the oven until most of the moisture has been released.

Lilac Oysterling

- cap is lilac
- grows on decaying hardwood
- gills extend onto stem
- stem is off-center or lateral
- spore deposit is white

DESCRIPTION: The **CAP** is initially broadly convex with edges that are inrolled. As it matures, the cap flattens, develops a central depression, and often becomes lobed or scalloped with wavy edges. At this point the fruiting body may resemble a funnel or be deeply vase-shaped. When young, the cap is violet to lilac-brown. Over time it becomes ocher-brown, yellow-brown, or tan, although the edges often retain some purple coloration. The young cap may have a felted texture. In maturity it often exhibits radial wrinkles, then develops cracks or small appressed scales in age. The **FLESH** within the cap is firm, tough, leathery, and white. It does not discolor when cut.

The **UNDERSIDE** of the cap has gills that extend down onto the stem. The gills are narrow and fork near the stem. Short gills ("lamellulae") are also present; they start at the edge and run only part of the way towards the center. The gills are initially whitish, tan or purplish, then fade to cream or pale brown. The **SPORE** deposit is white. The **STEM** is solid (not hollow) and is attached to the cap either laterally, off-center, or (rarely) at the center. It is roughly equal in thickness throughout its length, or may narrow downward. No veil or ring is present.

SIZE: Cap is 4–12(17) cm wide and 1–6(9) cm long; stem is 2–6(8) cm long and 1–2(3) cm thick.

LOOK-ALIKES: A number of inedible or poisonous species of *Clitocybe* are similar in appearance but differ in that they are terrestrial, rather than growing from wood. The edible but uncommon *Lentinus strigosus* is similar in appearance but differs in that it exhibits tufts of erect hairs on the cap. See also *Panellus serotinus (p. 52)*.

OCCURRENCE: In the spring, summer, fall and winter, this mushroom appears on decaying wood. It grows either alone or in clusters of several. Look for it on rotting hardwood, including fallen sticks and stumps, particularly in coastal areas.

COMMENTS: This mushroom has a mild flavor that is slightly bitter and reminiscent of turnips. Its odor is faintly earthy or like aniseed. For the western palate it is often regarded as inedible due to its toughness. Even so, it is considered a choice edible in France. Young specimens are more tender are therefore preferred.

Angel Wings

syn. Sugihiratake

- cap is ivory white and translucent
- flesh is extremely thin
- grows on decaying conifers
- spore deposit is white
- warning: deadly for kidney patients

DESCRIPTION: This pristinely white mushroom forms shelves or grows in overlapping clusters on decaying conifers. The **CAP** is initially convex and inrolled at the edges. In maturity it expands to become petal-, fan-, or tongue-shaped and is sometimes lobed. Pressure from neighbors in the same cluster can cause the edges to bend irregularly. The cap surface is characteristically shiny and ivory-white. It is slightly translucent and becomes creamy-white when old. The **FLESH** within the cap is pliant and characteristically very thin.

The **UNDERSIDE** of the cap has crowded, narrow gills that are ivory white when young, soon becoming creamy white or slightly yellowish. The gills run all the way to the base, which is a lateral point of attachment to the host tree. The **SPORE** deposit is white. There is no **STEM**; the base is white.

SIZE: Cap is 2–10 cm broad.

LOOK-ALIKES: Various species of *Crepidotus*, *Hohenbuehelia*, and *Panellus* are similar in appearance but are inedible or of unknown edibility. They differ most obviously in that they are relatively small (under 6 cm broad) and not generally as white. The edible *Pleurotus populinus* differs in that the cap is whitish but not ivory white, the flesh is thicker, and it fruits in the spring on **aspens** and **cottonwoods**. See also *Pleurotus ostreatus (p. 51)*.

OCCURRENCE: In the fall this species appears in dense clusters that form overlapping shelves. It grows on decaying conifer logs and stumps, on dead branches and on partly buried wood, including **Douglas-fir** and particularly **hemlock**.

COMMENTS: This mushroom is deadly poisonous for individuals with weakened kidneys. In other respects, it is a good edible with a mild and pleasant flavor and odor that give it culinary versatility. It is a popular mushroom in Japan.

Toothed Jelly Fungus

syn. White Jelly Mushroom/Fungus,
False Hedgehog, Cat's Tongue

- flesh is rubbery and translucent
- underside is covered in spines
- grows on conifer logs, stumps, debris

DESCRIPTION: This mushroom resembles a white, gelatinous tongue. It has a variable form that ranges from being fan- or kidney-shaped to being spoon- or bracket-like. When young it has a **HEAD** that is broadly convex with edges that are tucked-under. The head is translucent white or greyish white in color and sometimes becomes brownish in the center as it ages. The surface texture is smooth, gelatinous, finely fuzzy, and is not slimy. The flesh is translucent and rubbery.

The **UNDERSIDE** is covered in tiny, crowded, bluntly conical spines that constitute the spore-bearing surface. The color of the undersurface ranges from white to pale grey or faintly bluish. The spines extend partly down onto the stem. The **SPORE** deposit is white. The **STEM** is solid (not hollow). It is continuous with the head and tapers downwards. The surface is gelatinous, smooth, and similar in color to the head or paler. When the fruiting body grows from the top of a log or buried wood, which is most typical, the stem is well-developed, off-center and vertical. It elevates the cap to be almost as high as it is wide. When the fruiting body grows from the side of a host the stem is lateral and not well-developed.

SIZE: Cap is 2–7 cm broad and ½–1 cm thick; stem is up to 6 cm long and up to 1½ cm thick at the base; spines are under 3(5) mm long.

LOOK-ALIKES: None.

OCCURRENCE: In the late summer and early winter this fungus sometimes appears alone but more often in clusters. It favors damp, shady places and grows on well-decayed conifer logs, dead stumps, or on the terrestrial woody debris of conifers. In rare cases it can be found growing on standing trees. It is widely distributed.

COMMENTS: Some people consider this mushroom too bland and rubbery to be a good edible, whereas others eat it raw, marinated, or candied as a sugary treat. The process of candying involves boiling the mushroom and then steeping it in increasingly strong sugar solutions before drying off any remaining water. The high sugar saturation prevents spoilage.

Lion's Mane

syn. Satyr's Beard, Bearded Tooth, Monkey Head, Bearded Hedgehog, Pom Pom Mushroom

- resembles a white pom-pom
- spines are relatively long (1–7 cm)
- interior is unbranched
- grows on hardwood trees

DESCRIPTION: The fruiting body is a bulbous mass of tissue densely covered in downward-hanging, overlapping spines, or "teeth," making it somewhat reminiscent of a lion's mane or a pom-pom. When young the **OUTER SURFACE** is white or cream, then becomes slightly yellowish or pale brown as it ages. The **INTERIOR** is compact and unbranched or rarely has a few stout, basal branches. The **FLESH** within is firm, spongy, whitish, and does not discolor when cut.

The spines constitute the spore-producing surface. They are densely crowded together and on mature specimens are long enough to overlap and cascade. The spines are soft, slender, and have sharply pointed tips. The **SPORE** deposit is white. There is no **STEM**. The base of the fruiting body attaches to the host tree and is hidden from view. It is composed of tough, white tissue.

SIZE: Fruiting body is 10–20(40) cm broad; spines are 1–5(7) cm long.

LOOK-ALIKES: A number of inedible coral mushrooms are superficially similar in that they are white and spindly. They differ in that the spines do not hang downwards. The edible *Hericium coralloides* has downward-hanging spines but differs in that the fruiting body is branched, which gives it a more irregular shape that resembles a white sea coral. See also *Hericium abietis (p. 57)*.

OCCURRENCE: In the late summer and fall this fungus occasionally appears as a solitary growth from the wounds of trees, often high off the ground. It grows on both living and dead hardwood trees, including standing trunks, fallen logs, and large diameter branches, where it feeds off the central deadwood. It occurs at sites where old trees have been felled, sometimes fruiting from the cut-ends of logs.

COMMENTS: Considered a good edible, this fungus has a mild, pleasant odor and flavor that is reminiscent of crab-meat or lobster. This species has been the subject of research for its medicinal properties with respect to neural regeneration and cognitive heath. It is a commonly cultivated species that is dried and packaged for sale and is also sold in capsule form in health food stores.

Bear's Head

syn. Conifer Coral Hericium

- spines are pinkish/brown
- spines are under 1 cm long
- interior is branched
- grows on dead wood of conifers

DESCRIPTION: The fruiting body is a repeatedly branching structure that is covered with clusters of spines or "teeth." It is white when young, then becomes pinkish and finally ocher/brown when old. In maturity, the **OUTER SURFACE** has the appearance of small pom-poms irregularly clumped together. In some cases this fungus reaches massive proportions – among the heaviest on record is a specimen weighing over 100 pounds.

The **BRANCHES** are short and arise from a common base. Clusters of down-hanging spines are attached along the branches and also drape the branch tips, partly obscuring the branching nature of the fruiting body. The spines, which constitute the spore-bearing surface, are fleshy and brittle. The **FLESH** within the branches is whitish and does not discolor when cut. The **SPORE** deposit is white. At the base of the fruiting body is a thick, knob-like **STEM** composed of tough, white tissue.

SIZE: Fruiting body is 10–75 cm broad and similarly tall; main branches are up to 3 cm thick; spines are ½ to 1(2½) cm long.

LOOK-ALIKES: A number of inedible coral mushrooms are superficially similar in that they are white and have a spindly, branching structure. They differ in that the spines do not hang downwards. The edible *Hericium coralloides* has downward-hanging spines and a similar appearance but differs in that it lacks pinkish tones and the branches are less obscured by the spines, giving it a shape that more closely resembles a white sea coral. See also *Hericium erinaceus (p. 56)*.

OCCURRENCE: In the late summer and fall, this fungus appears either growing alone or occasionally as several together on dead conifer wood. Sometimes it occurs in the same place for several years. It can be found in old-growth forests, growing on stumps, decaying logs, and standing trunks of conifers, including **firs** and **hemlock**.

COMMENTS: This chewy mushroom is a fine edible. It has a mild flavor that is slightly nutty or fishy. Younger specimens are best for cooking, otherwise use low temperatures and extended cooking times to reduce chewiness. When harvesting, use a knife to cut away the main body and leave the stem base behind to promote subsequent fruiting.

Witch's Butter

syn. Yellow Brain,
Golden Jelly Fungus, Yellow Trembler

- resembles yellow/orange blobs
- grows on logs and branches
- grows in convoluted clusters
- clusters are 5-10 cm broad

DESCRIPTION: The fruiting bodies are irregularly-shaped, jelly-like blobs that form a convoluted cluster, sometimes resembling a brain. The color ranges from yellow to yellow-orange when conditions are normal to moist. When dried out, the fruiting bodies blacken and become bone-hard. No **STEM** is present. The **SPORE** deposit is white.

SIZE: Individuals are 5–8 mm long and 3–7 mm wide; clusters are 2–5(10) cm broad.

LOOK-ALIKES: The name **witch's butter** also refers to a number of similar-looking edible species, some of which can only be distinguished microscopically (e.g., *Tremella mesenterella*). The edible *Dacrymyces palmatus* and *Dacrymyces chrysospermus* are known as **orange witch's butter** and **orange jelly**. They differ from *Tremella mesenterica* in that the "blobs" are usually more orangey, have a whitish attachment point, and produce yellow spores. They grow on decaying conifer logs and sticks year-round, usually through cracks in the bark or on areas without bark. A number of other species of *Dacrymyces*, generally of unknown edibility, differ in being much smaller (under 4 mm broad).

The jelly-cup fungus *Guepiniopsis alpinus* has unknown edibility. The individual fruiting bodies differ from those of *Tremella mesenterica* in that they are more scattered, top-shaped or cup-shaped, and each hangs from a single point of attachment. *Heterotextus luteus*, of unknown edibility, differs in that the fruiting bodies have short, stout stems and a more scattered habit of growth. The edible *Tremella aurantia* differs in that it is somewhat larger and fruits on top of the fungus *Stereum hirsutum*, also known as **false turkey tail**.

OCCURRENCE: This mushroom is found year-round and typically appears on decaying hardwood, particularly fallen logs and branches that still retain bark. It is widespread and common in the Pacific Northwest.

COMMENTS: Although it lacks any flavor or odor and is not particularly appetizing, this species is edible when thoroughly cooked, such as when boiled or steamed. When under cooked it may cause digestive upset, and therefore should not be sautéed.

Jelly Ear

syn. Wood Ear,
Judas' Ear, Tree Ear

- grows on decaying trees
- flesh is rubbery/gelatinous
- surface is irregularly ribbed/veined

DESCRIPTION: This gelatinous fungus is irregularly cup-shaped and has ear-like in appearance, hence the common name **jelly ear**. In maturity the edges often become lobed and wavy. The color on both the inside and outside of the cup ranges from tan to brown. The **INSIDE** of the cup is smoothly wrinkled and exhibits ribs and branching veins. Its surface is covered in a very fine, velvety down and may have a white-frosted appearance.

The **OUTSIDE** of the cup is the spore-bearing surface. It is similar in color to the inside and shares the gelatinous consistency, wrinkled surface and whitish cast. The **FLESH** is thin, cartilaginous, and somewhat translucent when moist. When dried out it becomes tough and nearly black. The **STEM** is little more than a gathering together of the cup at the point of attachment to the host. It is attached laterally or sometimes centrally. The **SPORE** deposit is white.

SIZE: Fruiting body is 2–10(15) cm broad.

LOOK-ALIKES: The edible *Tremella foliacea* differs in that it grows in tightly packed, leafy clusters and only rarely occurs on conifers. The edible *Exidia recisa* differs in that it is smaller (up to 4 cm broad) and is brown to purple-brown. A number of terrestrial cup fungi are superficially similar and are not generally considered edible. They differ in that they do not grow on wood.

OCCURRENCE: This uncommon fungus appears in the late summer and fall shortly after rainfall. It can also be found early in the spring as a rehydrated holdover from the year before, or growing at higher elevations near melting snow. It grows either alone or in overlapping, shelving clusters that appear on living, damaged, or decaying parts of trees, including logs and fallen branches of both conifers and hardwoods. It prefers damp, shady locations.

COMMENTS: Although this mushroom has no odor or distinctive flavor, it does have an interesting, rubbery texture. The cultivated variety is very popular in eastern cuisines. Young specimens are considered best for cooking and may be used as a filler in mixed mushroom dishes. If preserved by drying, these mushrooms reconstitute well.

Apricot Jelly

syn. Red Jelly Fungus

- resembles a wavy tongue
- flesh is rubbery, translucent
- grows 5-15 cm tall

DESCRIPTION: This curious-looking fungus has a variable form. It is often tongue-shaped with wavy, lobed edges. Alternatively, it may be funnel-shaped and split on one side, giving it the appearance of a twisted cornet or a horn. In maturity its color ranges from whitish/pink to apricot/salmon, sometimes with purplish or reddish tones. When old it becomes more brownish.

The **UPPER** (inner) surface is smooth and sometimes frosted with a powdery bloom. The **OUTER** (under) surface is smooth in normal conditions and slightly sticky when wet. When old it develops veins or wrinkles. It constitutes the spore-bearing surface. The **SPORE** deposit is white. The flesh is flexible, rubbery and somewhat translucent. In the upper parts of the fruiting body it is soft, becoming progressively more cartilage-like near the short, cylindrical, **STEM**. The stem is off-center or lateral and is continuous with the rest of the fruiting body. It tapers downwards to a whitish base.

SIZE: Fruiting body is 10(18) cm broad and 5–15 cm tall; flesh is 2–3 mm thick; stem is up to 3 cm high.

LOOK-ALIKES: None.

OCCURRENCE: In the spring, summer or fall, this fungus appears on the ground after rainy periods, either alone or in small, dense, groups with each mushroom separated. It occurs in areas with standing conifers and acquires nutrients by breaking down well-rotted conifer wood that has become buried in the ground. It is widely distributed and common.

COMMENTS: This rubbery mushroom has a nondescript odor and a bland, indistinct flavor. Young specimens are sometimes used in salads or candied as a sugary treat. The process of candying involves boiling the mushroom and then steeping it in increasingly strong sugar solutions before drying off any remaining water. The high sugar saturation prevents spoilage. Older specimens become progressively tougher until they are inedible.

Club Coral

syn. Club Fungus, Truncated Club

- shape resembles a twister
- head is yellow when young
- top is flattish, depressed, or perforated
- grows 5–15 cm tall

DESCRIPTION: This club-shaped fungus is shaped like a tornado or twister. The relatively broad **HEAD** is flattened at the top and is continuous with the tapering body. The head is initially yellow, then becomes dull yellow to yellowish orange. In maturity it either remains flat or becomes centrally depressed, with the edges of the head becoming raised and irregularly bumpy. When old the head may rupture to reveal a hollow interior.

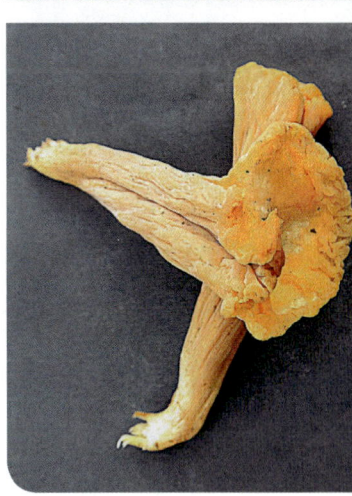

As it tapers, the **BODY** remains round in cross-section. Its color ranges from pinkish cinnamon to yellow brown or orange brown and it slowly discolors brown when bruised. Longitudinal grooves or wrinkles run down the sides and are smoothened away at the base. The **FLESH** is white to ocher and slowly discolors to brown when cut. It is soft when young and becomes spongier as it ages. The **BASE** of the fruiting body is cream or pale orange and is loosely matted with whitish, hair-like filaments. The **SPORE** deposit is white to pale yellow.

SIZE: Head is 3–8 cm broad, base is up to 1½ cm broad; fruiting body is 5–15 cm tall.

LOOK-ALIKES: The unpalatable *Clavariadelphus caespitosus* differs in that the head is rounded, the body has reddish tones, and it grows in tight, mat-forming clusters where the bases are often joined. The unpalatable *Clavariadelphus occidentalis* differs in that the head is rounded and the sides are flattened. A number of other inedible species, such as *Clavariadelphus subfastigiatus* and *Clavariadelphus ligula* are somewhat similar in appearance, but differ in that the head is rounded, not truncated.

OCCURRENCE: In the late summer and fall this common mushroom appears either scattered or in groups of a few on the ground, with each mushroom separated and not forming a mat. It can be found growing in duff or needle litter under conifers.

COMMENTS: This fungus has a sweet or bittersweet flavor and a faint but pleasant odor. Sweet varieties can be sautéed and eaten as a dessert. Some consider it to be a choice edible; however, it may have a mildly laxative effect.

Washington State Personal Use Mushroom Harvesting Rules

This section outlines the rules and regulations for **personal-use harvesting** in various districts of Washington. Mushrooms that are harvested for personal use cannot be sold, bartered, or given away. This section does not cover commercial mushroom harvesting rules. It also does not cover the mushroom harvesting rules for groups, except where noted.

According to Washington state law, a **Washington Specialized Forest Products Permit** must be obtained to harvest and transport more than five gallons of wild edible mushrooms anywhere in the state. One permit is required per vehicle. To obtain a permit contact a Forest Service office.

Additional regulations for each district are provided below. Note that personal-use harvesting is generally open year-round, with the exception of incidental closures or access restrictions.

1. Olympic National Park
2. Olympic National Forest
3. Mt. Baker Snoqualmie National Forest
4. North Cascades National Park
5. Okanogan National Forest
6. Wenatchee National Forest
7. Mount Rainier National Park
8. Gifford Pinchot National Forest
9. Umatilla National Forest
10. Colville-Kaniksu National Forest
11. Washington State Parks
12. Natural Area Preserves (NAP)
13. Bureau of Land Management (BLM) Land
14. Department of Natural Resources (DNR) Forest and Land Trusts AKA State Forests
15. Natural Resource Conservation Areas (NRCA)
16. National Wildlife Refuges
17. Department of Fish & Wildlife Lands
18. City Parks

1. OLYMPIC NATIONAL PARK

- Individuals do not require a park permit to harvest mushrooms.
- Daily limit is one gallon per species to a limit of three species per day (three gallon total).
- Chanterelle caps must be at least 1" in diameter.

2. OLYMPIC NATIONAL FOREST

- Individuals do not require a permit to harvest mushrooms.
- Daily limit is one gallon per species to a limit of three species per day (three gallon total).
- Chanterelle caps must be at least 1" in diameter.

3. MT. BAKER SNOQUALMIE NATIONAL FOREST

- Harvesting up to one gallon of mushrooms per day is allowed without a permit.
- If harvesting more than one gallon, a **Free-Use Permit** is required. This permit has no cost and allows gathering of up to five gallons per year per household. The **Free-Use Permit** is available at local ranger offices.
- If harvesting more than five gallons, a **Day Use Permit** is required, which varies in cost from $25-$125 depending on the duration of the permit (2 days – 6 months).

4. NORTH CASCADES NATIONAL PARK

- Individuals do not require a park permit to harvest mushrooms.
- The harvesting or possession of any combination of edible fruits, berries and/or mushrooms is limited to one liter (one quart) per person per day. Apples do not count towards the total and may be gathered in unlimited amounts for non-commercial use.

5. OKANOGAN NATIONAL FOREST

- Individuals are permitted to harvest up to five gallons of mushrooms per day.
- Harvesting any amount of morel mushrooms requires carrying an **Incidental Use Mushroom Information Sheet** (northernbushcraft.com/wa/is). The sheet is free and must be in possession of the forager while in the forest.
- Up-to-date information and maps of suggested foraging locations are available on the Okanogan-Wenatchee National Forest website (northernbushcraft.com/wa/onf).

6. WENATCHEE NATIONAL FOREST

- Individuals are permitted to harvest up to five gallons of mushrooms per day.
- Harvesting of any amount of morel mushrooms requires carrying an **Incidental Use Mushroom Information Sheet** (northernbushcraft.com/wa/is). The sheet is free and must be in possession of the forager while in the forest.
- Up-to-date information and maps of suggested foraging locations are available on the Okanogan-Wenatchee National Forest website (northernbushcraft.com/wa/onf).

7. MOUNT RAINIER NATIONAL PARK

- Individuals do not require a park permit to harvest mushrooms.
- Daily limit is one gallon of mushrooms per day.

8. GIFFORD PINCHOT NATIONAL FOREST

- Individuals are permitted to harvest up to two gallons of mushrooms per day for up to 10 days in calendar year (not necessarily consecutive), for a total of 20 gallons per year.
- Individuals are required to carry a **Free-Use Permit** and a **Harvest Area Map**. The permit and map are available at no cost at ranger stations or online at https://apps.fs.usda.gov/gp

9. UMATILLA NATIONAL FOREST

- Individuals are permitted to harvest up to five gallons of mushrooms without a permit.
- Harvesting more than five gallons requires individuals to purchase a permit, at a cost of $2/day, with a minimum term of 10 days.

10. COLVILLE-KANIKSU NATIONAL FOREST

A. NEWPORT, REPUBLIC, SULLIVAN LAKE & THREE RIVERS

- Individuals do not require a permit to harvest mushrooms.
- Daily limit is three gallons or 60 mushrooms, whichever is less in volume.
- Individuals are prohibited from harvesting pine/matsutake mushrooms.

B. PRIEST LAKE

- Harvesting up to 1 gallon of mushrooms per day is allowed without a permit.
- A **Free-Use Permit** is available to anyone over 12 years of age. The permit allows individuals to harvest up to five gallon of mushrooms per day with a yearly limit of 20 gallons.
- A **Free-Use Permit** can be obtained at a ranger office at no cost.

11. WASHINGTON STATE PARKS

- State parks allow individuals to harvest, possess, and transport edible plants, mushrooms, berries, and nuts up to a limit of two gallons per day, regardless of species, unless otherwise posted at the park. See **State Park Rules** (northernbushcraft.com/wa/spr).
- Groups are required to contact rangers at individual parks for rules and regulations.

12. NATURAL AREA PRESERVES (NAP)

Foraging of any kind is prohibited.

13. BUREAU OF LAND MANAGEMENT (BLM) LAND

Individuals do not require a permit to harvest special forest products, including grasses, seeds, roots, bark, berries, mosses, greenery (fern fronds, berries, etc.), mushrooms, tree seedlings, transplants, poles, posts, and firewood.

14. DEPARTMENT OF NATURAL RESOURCES (DNR) FOREST AND LAND TRUSTS AKA STATE FORESTS

To harvest mushrooms on DNR land, individuals are required to have a **Discover Pass** displayed on the vehicle, at the risk of a $99 fine. The **Discover Pass** costs $10 for a day pass and $30 for an annual pass. It can be purchased online at https://store.discoverpass.wa.gov, via phone at 866-320-9933, or at State Park offices. A list of vendors is available online at: http://discoverpass.wa.gov/133/Where-to-Buy

A. AHTANUM STATE FOREST

Individuals are required to have a **Discover Pass** to harvest mushrooms.

B. BLANCHARD STATE FOREST

- Individuals are required to have a **Discover Pass** to harvest mushrooms.
- Annual limit is five gallons per person.

C. CAPITOL STATE FOREST

- Individuals are required to have a **Discover Pass** to harvest mushrooms.
- Daily limit is three gallons of a single mushroom species per day, not to exceed nine gallons per mushroom per year.
- Groups must contact the Recreation Manager to be assessed on a case by case basis, and groups of over 25 people are required to have an insurance bond.

D. ELBE HILLS AND TAHOMA STATE FOREST, GREEN MOUNTAINS AND TAHUYA STATE FOREST, TIGER MOUNTAIN STATE FOREST & YACOLT BURN STATE FOREST

- Individuals are required to have a **Discover Pass** to harvest mushrooms .
- Daily limit is three gallons of a single mushroom species per day.
- Annual limit is nine gallons per mushroom species per year.

E. LITTLE PEND OREILLE STATE FOREST, LOOMIS – LOUP STATE FOREST, OLYMPIC PENINSULA STATE FOREST & TEANAWAY STATE FOREST

- Individuals are required to have a **Discover Pass** to harvest mushrooms.
- Daily limit is five gallons of mushrooms per day.

15. NATURAL RESOURCE CONSERVATION AREAS (NRCA)

Foraging of any kind is prohibited.

16. NATIONAL WILDLIFE REFUGES

Foraging of any kind is prohibited.

17. WASHINGTON DEPARTMENT OF FISH & WILDLIFE (WDFW)LANDS

- Individuals do not require a permit to harvest mushrooms.
- Some areas are restricted to hunting and foraging.
- GoHunt interactive online map shows WDFW lands, boundaries, restricted areas, and management areas: http://apps.wdfw.wa.gov/gohunt/

18. CITY PARKS

Check mushroom harvesting rules with your municipality.

A. SEATTLE

Mushroom picking is not specifically prohibited in city parks, however all other types of foraging are prohibited (**Seattle Municipal Code 18.12.070B**).

B. SPOKANE

Mushroom picking is not prohibited in city parks (https://my.spokanecity.org/parks/rules).

C. TACOMA

Removal of mushrooms is prohibited in city parks (**Tacoma Municipal Code 8.27.100**).

D. VANCOUVER

Mushroom picking is not specifically prohibited in city parks, however all other types of forging are prohibited (**Vancouver Municipal Code 15.04.040**).

E. BELLEVUE

The removal of any vegetation is prohibited (**Bellevue Municipal Code 3.43.335**).

F. KENT

Mushroom picking is not specifically prohibited in city parks, however all other types of foraging are prohibited (**Kent Municipal Code 4.01.020**).

acrid: intensely and unpleasantly strong or sharp in taste or smell.

adnate: referring to gills that are broadly attached to the stem.

alpine: living or growing above the timber line.

adnexed: referring to gills that are narrowly attached to the stem.

annulus: remnant of the partial veil that in mature mushrooms surrounds the lower part of the stem.

appressed-fibrillose: having fibrils that are pressed close or lying flat.

appressed-scaly: having flattened scales that are pressed close.

areolate: having a surface cracked or divided into somewhat polygonal sections, like dried-cracked mud.

ascomycete: any fungus of the class Ascomycetes (or subdivision Ascomycota) in which the spores are formed inside an ascus.

basidiomycete: any of various fungi of the subdivision Basidiomycota.

bloom: a powdery deposit on a surface.

bolete: a general name for a stemmed fleshy mushroom with pores rather than gills on the underside of the cap.

bracket fungus: a general term for a woody fungus that forms a shelflike fruitbody, commonly called a "conk."

buff: a medium to dark tan color.

bulbous: having a base that is bulb-like or swollen.

button: a young mushroom that resembles a button.

caespitose: term used to describe a cluster of fruiting bodies having stems that grow densely together but not fused.

cap: a fruiting structure resembling an umbrella that forms the top of a stemmed fleshy fungus such as a mushroom, also called the pileus.

cartilaginous: tough and firm but pliable.

clavate: club-shaped.

close (gills): referring to gills that almost touch their neighbor but maintain a visible space between.

concolorous: with multiple parts of a fruiting body being of the same color.

conifer: any gymnospermous tree or shrub bearing cone.

convex: curving or bulging outward.

corrugated: shaped into alternating parallel grooves and ridges as wrinkles.

crenate: having an edge with small rounded teeth.

crenulate: finely crenate.

crossveins: interconnections between neighboring gills or ridges of a hymenium.

crowded: referring to gills that are crammed together with little space between.

cuticle: the outer layer of tissue covering the exterior of the cap or stem.

deciduous: shedding foliage at the end of the growing season.

decurrent: referring to gills that extend down the stem.

deliquescing: becoming reduced to a liquid.

depressed: referring to a cap having a sunken center.

disc: referring to the central area of the cap directly above the attachment of the stem.

eccentric: referring to a stem that is off-center but not on the edge of the cap.

equal: having an unchanging stem diameter throughout the entire length.

evanescent: briefly present before disappearing.

fairy ring: a ring or arch of fungi marking the periphery of the perennial underground growth of the mycelium.

farinaceous: having the taste or odor of freshly ground flour; resembling starch, covered with fine particles.

fibril: a very slender, minute, thread-like fiber.

fibrillose: have a surface covered with delicate fibrils.

fibrillose-matted: having a surface covered in interwoven fibrils that give a felted texture.

fibrillose-scaly: having a surface covered in scales made up of fibrils.

flesh: sterile (non-spore producing) tissue of a fruiting body, also called the context.

floccose: having tufts of soft, loose, woolly scales.

foray: an organized group excursion into the forest to look for fungi.

free: referring to gills that do not reach the stem.

fruiting body: the organ of a fungus that produces spores, often regarded as a mushroom.

fungus: an organism of the kingdom Fungi lacking chlorophyll and feeding on organic matter.

gasteromycete: any fungus of the class Gasteromycetes (e.g., puffballs).

genus: taxonomic group containing one or more species.

gill: any of the radiating blade-like spore-producing structures on the underside of the cap of a mushroom or similar fungus, also called lamella (lamellae pl.)

glabrous: having no hair or similar growth; smooth, bald.

gleba: fleshy spore-bearing inner mass of a gasteromycete (e.g., a puffball).

globose: spherical, globe-shaped.

glutinous: covered with a sticky substance.

gregarious: tending to form a group in a small area but each separated.

hardwood: the wood of broad-leaved dicotyledonous trees (as distinguished from the wood of conifers).

hygrophanous: changing color or transparency in mushroom tissue as it loses/absorbs water.

hymenium: spore-bearing layer of cells in certain fungi.

hypha (hyphae pl.): individual strand of a mycelium.

infundibuliform: funnel shaped.

inrolled: rolled inward.

involute: see inrolled.

kingdom: the highest taxonomic group into which organisms are grouped; one of five biological categories: Monera or Protoctista or Plantae or Fungi or Animalia.

laciniated: referring to a surface that is divided into deep, narrow, irregular segments.

lamellula (lamellulae pl.): a short mushroom gill between normal gills (lamellae) that does not extend all the way from the edge of the cap to the stem.

lignicolus: living on or in wood.

margin: area constituting the edge of the cap.

membranous: characterized by the formation of a membrane of thin skin.

mushroom: fruiting body of any of numerous edible fungi.

mycelium (mycelia pl.): the vegetative part of a fungus consisting of a mass of branching threadlike hyphae.

mycology: the branch of botany that studies fungi.

mycorrhiza: a fungus that grows in mutually beneficial association with the root system of a vascular plant.

notched: narrowly attached gills that have a notch-like region where they meet the stem.

ochraceous: colored ocher (dull brownish yellow).

oliveaceous: colored olive green.

ovate: egg-shaped with the broader end at the base.

ovoid: oval-shaped.

pallid: deficient or lacking in color.

parasitic: referring to a fungus that acquires its nutrients from a living organism.

partial veil: membrane of the young sporophore of various mushrooms extending from the margin of the cap to the stem and is ruptured by growth; represented in mature mushroom by an annulus around the stem and sometimes a cortina on the margin of the cap.

pellicle: the peelable skin of a mushroom cap.

perennial: a fruitbody that recurs and produces spores for more than one year.

peridium: outer layer of the spore-bearing organ in many fungi.

phylum: the major taxonomic group of animals and plants; contains classes.

pileus: a fruiting structure resembling an umbrella or a cone that forms the top of a stemmed mushroom.

polypore: woody pore fungi; any fungus of the family Polyporaceae or family Boletaceae having the spore-bearing surface within tubes or pores; the fruiting bodies are usually woody at maturity and persistent

pore: opening of a tube on boletes and polypores.

poroid: a fertile surface having circular openings where spores are produced.

pruinose: appearing to be covered with a fine, powdery granules; frosted.

punctate: referring to a surface that is studded with dots or tiny holes.

recurved: bent backwards.

reticulate: having a net-like pattern.

rhizoid: any of various slender filaments that function as roots in fungi.

rhizomorph: a dense mass of hyphae forming a root-like structure of many fungi.

ring: see annulus.

rivulose: marked by thin, winding lines, like the rivers on a map.

rugulose: referring to a surface that is uneven because it is wrinkled or corrugated.

russet: colored reddish brown.

saprobe: a fungus that lives in and derives its nourishment from decaying organic matter.

scabers: small, roughened projections that occur on the stem of some boletes, especially in the genus *Leccinum*.

scabrous: referring to a stem covered in scabers.

scale: a flaky, plate-like, or delineated piece of tissue.

scaley: covered in scales.

scattered: refers to habit of growth where fruiting bodies are widely separated (30-60cm apart).

scurfy: coarsely granular.

seceding: referring to gills that were attached to the stem but pull away from it.

sessile: lacking a stem.

short-decurrent: refers to gills that extend only slightly down the stem.

sinuate: referring to gills that have a concave indentation near the stem.

solitary: referring to a habit of growth where the fruiting body occurs alone.

spathulate: shaped like a spatula; oval with a narrowed base.

species: taxonomic group of similar organisms whose members are capable of interbreeding.

sphagnum: any of various pale or ashy mosses of the genus Sphagnum whose decomposed remains form peat.

spore: a microscopic, reproductive body produced by a fungus that develops into a new individual.

spore case: structure of a gasteromycete holding the spore mass.

spore deposit: a mass deposition of spores obtained when the spores of fruitbody are allowed to fall onto a surface beneath.

stem: the structure that supports the cap or head of a fruitbody, also called a stipe.

sterile: tissue incapable of reproducing by creating spores.

stipe: see stem.

striate: marked with striations (lines or grooves).

subdecurrent: referring to gills that run parallel to the cap for most of their length but have edges that curve down onto the stem at the point of attachment.

subdistant: referring to gills that are intermediate between close and widely spaced.

subglobose: nearly globose.

subspecies: a taxonomic group that is a division of a species; usually arises as a consequence of geographical isolation within a species.

substrate: material on or within which a fungus grows.

sulcate: having lines more deeply furrowed than striate.

teeth: spine-like, spore-bearing structures that occur on some fungi.

terrestrial: growing on the ground.

tomentose: densely covered with short matted woolly hairs.

umbilicate: referring to a cap with a central perforation into the stem.

veil: a membranous covering attached to the immature fruiting body of certain mushrooms.

ventricose: referring to a stipe that is narrow at the base and the apex but broader in the middle.

volva: cuplike structure around the base of the stem of certain fungi.

EXTERNAL RESOURCES

Mushroom Observer
http://mushroomobserver.org

Facebook's Pacific Northwest Mushroom Identification Forum
https://www.facebook.com/groups/PNWMushroomID

Facebook's Mushroom Identification Forum
https://www.facebook.com/groups/MushroomID/

Puget Sound Mycological Society
http://www.psms.org

South Sound Mushroom Club
http://www.southsoundmushroomclub.com/

Mycological Society of America
http://www.msafungi.org

North American Mycological Association
https://www.namyco.org/

Pacific Northwest Key Council - Keys to Mushrooms of the Pacific Northwest
http://www.svims.ca/council/keys.htm

Modern Forager - Web-based Burn Maps for Morel Hunting (product)
https://www.modern-forager.com/burn-morels/

REFERENCES

Ammirati, Joseph F. *Poisonous Mushrooms of the Northern United States and Canada* (U of Minnesota Press, 1985)

Bessette, Alan. *Mushrooms of North America in Color: A Field Guide Companion to Seldom-illustrated Fungi* (Syracuse University Press, 1995)

Bessette, Alan, William C. Roody and Arleen Rainis Bessette. *North American Boletes: A Color Guide to the Fleshy Pored Mushrooms* (Syracuse Uni. Press, 2000)

Beug, Michael, Alan E. Bessette and Arleen R. Bessette. *Ascomycete Fungi of North America: A Mushroom Reference Guide* (University of Texas Press, 2014)

Christensen, Clyde Martin. *Edible Mushrooms* (U of Minnesota Press, 1964)

Coker, William Chambers. *The Club and Coral Mushrooms (Clavarias) of the United States and Canada* (Courier Corp, 1923)

Davis, Mike, Robert Sommer and John Menge. *Field Guide to Mushrooms of Western North America* (University of California Press, 2012)

Fischer, David W. and Alan E. Bessette. *Edible Wild Mushrooms of North America: A Field-to-kitchen Guide* (University of Texas Press, 2010)

Holmberg, Pelle and Hans Marklund. *The Pocket Guide to Wild Mushrooms: Helpful Tips for Mushrooming in the Field* (Skyhorse Publishing Inc., 2013)

Jones, Bill. *The Deerholme Foraging Book: Wild Foods from the Pacific Northwest* (TouchWood Editions, 2014)

Kuhnlein, Harriet V. and Nancy J. Turner. *Traditional Plant Foods of Canadian Indigenous Peoples: Nutrition, Botany, and Use* (Taylor & Francis, 1991)

Lincoff, Gary. *The Complete Mushroom Hunter: An Illustrated Guide to Finding, Harvesting, and Enjoying Wild Mushrooms* (Quarry Books, 2011)

Marley, Greg A. *Chanterelle Dreams, Amanita Nightmares: The Love, Lore, and Mystique of Mushrooms* (Chelsea Green Publishing, 2010)

McKnight, Kent H. and Vera B. McKnight. *Field Guide to Mushrooms: North America* (Houghton Mifflin Harcourt, 1998)

Meuninck, Jim. *Basic Illustrated Edible and Medicinal Mushrooms* (Rowman & Littlefield, 2015)

Miller, Orson K. and Hope Miller. *North American Mushrooms: A Field Guide to Edible and Inedible Fungi* (Falcon Guide, 2006)

Phillips, Roger. *Mushrooms and Other Fungi of North America* (Firefly Books, 2010)

Rogers, Robert. *The Fungal Pharmacy: The Complete Guide to Medicinal Mushrooms and Lichens of North America* (North Atlantic Books, 2012)

Schalkwijk-Barendsen, Helene M. *Mushrooms of Northwest North America* (Lone Pine, 1994)

Schwab, Alexander. *Mushrooming Without Fear: The Beginner's Guide to Collecting Safe and Delicious Mushrooms* (Skyhorse Publishing, Inc., 2007)

Siegel, Noah and Christian Schwarz. Mushrooms of the Redwood Coast: A Comprehensive Guide to the Fungi of Coastal Northern California (Potter/TenSpeed/Harmony, 2016)

Schalkwyk, Helene M. E. *Mushrooms of Western Canada* (Lone Pine, 1991)

Trudell, Steve and Joseph F. Ammirati. *Mushrooms of the Pacific Northwest* (Timber Press, 2009)

Winkler, Daniel. *A Field Guide to Edible Mushrooms of the Pacific Northwest* (Harbour Publishing Company, Limited, 2011)

PHOTO CREDITS

The following works are used by permission of their respective copyright holders.

Leah Bendlin: 21b; **Tom Cervenka:** vi, xix, xxi-f, 2a, 5a, 5b, 8a, 9a, 9b, 13a, 13b, 14a, 18a, 23c, 25a, 25b, 25c, 26a, 27a, 27b, 27c, 30a, 30b, 30c, 34b, 42b, 43b, 49a, 49b, 49c, 52a, 52b, 52c, 55a, 55b, 58b, 58c; **Darvin DeShazer:** 4a, 4b, 4c, 23b, 36c, 36b, 44b, 53c; **Johannes Harnisch:** 20a; **May Kald:** 41a; **Guy Kennedy:** 32b; **Krista Lynn Farris:** 11c; **Duane Neuens:** 41b; **Robert Parcher:** 32a; **Ron Pastorino:** 2b, 3b, 7a, 7b, 10b, 11a, 11b, 12a, 20b, 21a, 28b, 29b, 38a, 40a, 40c, 44a, 45a, 45b, 48a, 53a, 53b; **Alan Rockefeller:** vii, 6a, 35b, 38b, 38c, 39a, 39c, 40b, 48b, 58a; **Tim Sage:** 22a, 22c, 26b, 36a, 36b; **Jessica Williams:** 31b

The following works are licenced under the Creative Commons or are available in the public domain. Each entry gives the photographer name followed by photo page numbers and positions. The code "CC BY 2.0" refers to creativecommons.org/licenses/by/2.0/. The code "CC BY 2.5" refers to creativecommons.org/licenses/by/2.5/. The code "CC BY 3.0" refers to creativecommons.org/licenses/by/3.0/. The code "CC BY ND 2.0" refers to creativecommons.org/licenses/by-nd/2.0/. The abbreviation "FL" refers to flickr.com, the abbreviation "WM" refers to Wikimedia Commons, and the abbreviation "PD" refers to works in the Pubic Domain.

Archenzo: xxi-a *[CC BY 3.0 via WM]*; **Benketaro:** 51b *[CC BY 2.0 via FLK]*; **Bill Bryden:** 19a [CC BY 2.0 via FLK]; **Courtney Celley/USFWS:** 24c *[CC BY 2.0 via FLK]*; **Alfred Crabtree:** 56a *[CC BY-ND 2.0 via FLK]*; **Dick Culbert:** xxi-b, 8b, 10a, 15a, 50c, 61b *[CC BY 2.0 via FLK]*; **Scott Darbey:** 3a, 6b, 34a, 47c, 57a, 57b *[CC BY 2.0 via FLK]*; username denAsuncioner: 42a *[CC BY-ND 2.0 via FLK]*; **Jenny Downing:** 59a *[CC BY 2.0 via FLK]*; username Gargoyle888: 50a *[CC BY 3.0 via WM]*; username Gljivarsko Drustvo Nis: xxi-i, 29a *[CC BY 2.0 via FLK]*; **Jean-Pol Grandmont:** 10c *[CC BY 3.0 via WM]*; **Andrew Green:** 31a *[CC BY 2.0 via FLK]*; **Gail Hampshire:** 16b, 22b *[CC BY 2.0 via FLK]*; **Lyanda Haupt:** 18b *[CC BY 2.0 via FLK]*; **Jason Hollinger:** xxi-e, xxi-g, 14b, 17a, 17b, 19c, 26c, 28a, 35a, 37a, 37b, 60a, 60c, 61a *[CC BY 2.0 via FLK]*; **Rocky Houghtby:** 33a *[CC BY 2.0 via FLK]*; username Hr.icio: 56b *[CC BY 3.0 via WM]*; **D. Kolegayev:** 47a, 47b *[CC BY 2.0 via FLK]*; **H. Krisp:** 46b, 60b *[CC BY 3.0 via WM]*; **Christine Majul:** 24b *[CC BY 2.0 via FLK]*; **Marian Maxwell:** 12b *[CC BY 2.0 via FL]*; **Jinx McCombs:** 46c *[CC BY-ND 2.0 via FLK]*; username Michael M.: 19b *[CC BY-ND 3.0 via FLK]*; **Ben Mitchell:** 54b [PD]; **Daniel Neal:** xxi-c, xxi-d, 51a [CC BY 2.0 via FLK]; **Tomasz Przechlewski:** 50b [CC BY 2.0 via FLK]; username Sasata: 43a [CC BY 3.0 via WM]; **Katja Schulz:** 16a, 46a [CC BY 2.0 via FLK]; **Bernard Spragg:** xxi-h [CC BY 2.0 via FLK]; **Julie Steiner:** 59b [CC BY 2.0 via FLK]; **Sheila Sund:** 54a [CC BY 2.0 via FLK]; **Sir John Tenniel:** 1 [PD via WM]; **Gerry Thomasen:** 34c [CC BY 2.0 via FLK]; **Jason Vanderhill:** 24a [CC BY 2.0 via FLK]; **Sarah Ward:** 59c [CC BY 2.0 via FLK]; **Tony Webster:** 23a *[CC BY-2.0 via FLK]*; **Becka York:** 33b [CC BY 2.0 via FLK];

72

INDEX

INDEX

ABOUT THE AUTHOR

Tom Cervenka is an avid naturalist and author whose northernbushcraft website has introduced over a million visitors to practical information on wild edible mushrooms, plants, and seashore edibles of the Pacific Northwest and Canada. Some of his other books include "Wild Edible Mushrooms of British Columbia," "Wild Edible Mushrooms of Alberta" and "Wild Edible Mushrooms of Ontario." Tom is an alumnus of the University of Alberta and a member of the Vancouver Mycological Society.

Made in the USA
Monee, IL
26 September 2019